# Climate Change and Urban Settlements

Climate change and urbanization are two of the greatest challenges facing humanity in the 21st century, and their effects are converging in dangerous ways. Cities contribute significantly to global warming, and as the world further takes a rural–urban population tilt, the next few decades pose a great challenge in addressing global disparities in the access and allocation of carbon.

This book explores the ways in which cities, through their spatial development, contribute to greenhouse gas (GHG) emissions and looks at the ways in which rapidly urbanizing cities in low- and middle-income countries can be planned to reduce overall GHG emissions. The book considers key questions, such as: What should be the appropriate economies of scale for cities in a country? What is the most favourable rate of urbanization? What should be the most suitable spatial pattern for a city? And what are appropriate regulatory, economic or governance mechanisms to achieve a low-carbon society? These issues are explored through data analysis of over 156 developing countries and through a specific case study of India. India acts as an interesting example of how societies undergoing rural-to-urban transformations could become green within the planetary boundaries while systematically addressing national and local urban governance. The research concludes with a future pathway that is committed to low-carbon and high-equity spatial development, and will find pertinence to researchers and practitioners alike.

This book provides a new tool for policymakers, planners and scholars to rationally and equitably account for global carbon space, prioritize low-carbon strategies for national urbanization and planning individual cities, in addition to recommending an urban governance framework inclusive of green agenda.

**Mahendra Sethi** is an urban environment expert and recipient of the United Nations University – Institute for the Advanced Study of Sustainability (UNU-IAS) PhD Fellowship at UNU-IAS, Japan. His field of research explores the role of cities at the interface of global environmental change with a particular focus on developing countries. In India, he is associated with the National Institute of Urban Affairs, as editor of the research journal *Urban India*.

## Routledge Studies in Hazards, Disaster Risk and Climate Change

Series Editor: Ilan Kelman, Reader in Risk, Resilience and Global Health at the Institute for Risk and Disaster Reduction (IRDR) and the Institute for Global Health (IGH), University College London (UCL)

This series provides a forum for original and vibrant research. It offers contributions from each of these communities, as well as innovative titles that examine the links between hazards, disasters and climate change, to bring these schools of thought closer together. This series promotes interdisciplinary scholarly work that is empirically and theoretically informed, with titles reflecting the wealth of research being undertaken in these diverse and exciting fields.

### Published

# Climate Change and Urban Settlements

## A Spatial Perspective of Carbon Footprint and Beyond

**Mahendra Sethi**

Routledge
Taylor & Francis Group

LONDON AND NEW YORK

First published 2017 by Routledge

2 Park Square, Milton Park, Abingdon, Oxfordshire OX14 4RN
52 Vanderbilt Avenue, New York, NY 10017

*Routledge is an imprint of the Taylor & Francis Group, an informa business*

First issued in paperback 2018

*British Library Cataloguing in Publication Data*
A catalogue record for this book is available from the British Library

*Library of Congress Cataloging in Publication Data*
Names: Sethi, Mahendra, author.
Title: Climate change and urban settlements: a spatial perspective of
    carbon footprint and beyond / Mahendra Sethi.
Description: Abingdon, Oxon; New York, NY: Routledge, 2017. |
Series: Routledge studies in hazards, disaster risk and climate change |
    Includes bibliographical references and index.
Identifiers: LCCN 2016048021| ISBN 9781138226005 (hbk) | ISBN
    9781315398501 (ebk)
Subjects: LCSH: City planning – Climatic factors. | Carbon dioxide
    mitigation. | Sustainable urban development.
Classification: LCC TD168.5 .S47 2017 | DDC 363.738/746091732 – dc23
LC record available at https://lccn.loc.gov/2016048021


ISBN: 978-1-138-22600-5 (hbk)
ISBN: 978-0-367-21893-5 (pbk)


Typeset in Times New Roman
by Swales & Willis Ltd, Exeter, Devon, UK

*This investigation is a small step in bringing global concerns such as climate change down to local priorities. As a public service, I dedicate this work to the people of my nation, the millions of hungry, impoverished, marginalized and homeless souls whose future is at the mercy of regular debates on environment and energy resources at the global level, while they stand as mute spectators. As thinkers, scientists, researchers, fellow countrymen and, above all, as human beings, we owe them their unclaimed due.*

# Contents

# Figures

# Tables

# Boxes

# Foreword

Climate change and urbanization are two of the greatest challenges currently facing humanity in the 21st century.

(UN-Habitat, 2011, p. 1)

Despite much talk about the importance of urbanization and climate change, little is known about the relations between them, scientifically. When I say 'scientifically', it means with clear methods, methodology and data to support the claims about the relations. Just to give an example, a few years back the United Nations Environment Programme (UNEP) and UN-Habitat mentioned that urban areas covered 2% of the Earth's surface and consumed more than 75% of its resources. However, when I tried to check how these numbers were estimated or calculated, I was surprised that they seemed, at best, like a guess. Therefore, we are talking all the time about urbanization but we do not yet know precisely how much urban we are. Indeed, we cannot even agree on "what is urban", as countries have different definitions. On the other hand, climate change, which has been studied systematically in the last there decades, has a large amount of scientific information thanks to the Intergovernmental Panel on Climate Change (IPCC). Consequently, the IPCC has been fundamental to understanding a complex chain of relations in the world's climate and influencing policymaking.

Thus, we have been studying urbanization, as a field of research, for more than 100 years, but we realize we know very little. One of the reasons is that we do not have an 'IPCC' for cities, though the complexity of urbanization is well known and its consequences for humanity widespread. Therefore, this book by Dr. Mahendra Sethi can make tremendous contributions to better comprehend urbanization and its non-obvious relations with climate change. He uses different spatial and non-spatial methods and data to elucidate the intricate relations between cities and climate change. The book is unique in its methodological approaches and longitudinal and geographical breadth for analyzing the city climate change phenomenon. The book also advances our understanding on how to fill the governance and policy gaps that prevent us from having more sustainable urbanization patterns, which are fundamental to tackle climate change.

The book includes a comprehensive analysis of urbanization processes in approximately 200 countries using a database of more than five decades. One key point is the correlation of urbanization rates with carbon dioxide emissions per capita, which is higher than the correlation between GDP per capita and carbon dioxide emissions per capita. This has two major implications. The first is that changes in our urbanization patterns are more important than we thought regarding climate change mitigation. The second implication is the gap existent between urbanized and mostly rural countries in carbon dioxide emissions per capita, as well as between rural–urban differences in the same country, particularly in developing countries such as India.

The new urban agenda coming out of Habitat III emphasizes inequalities are a key component for moving towards a more sustainable urban future. Leaders have the opportunity to change the game of inequality by addressing the core of it (i.e. the patterns of unsustainable urbanization in the North, and more recently in the South). Changes will take a different view on the role of cities and their main long-term development objectives, moving urban development from being based on uncontrolled consumption and concrete building to quality of life, resource conservation and a culture of sufficiency.

In sum, Dr. Sethi's contributions are far beyond the interpretations of patterns of urbanization, showing also the immense urban–rural inequalities in the emissions of carbon dioxide. This has important practical consequences for how we can provide evidence of this new dimension of inequalities, which adds up one more layer of complexity to the debates on the North–South historical carbon emission inequalities. Thus, the book is a must-read for scholars, professionals and students who want to have rigorous analyses of the relations between urbanization and climate change.

Jose A. Puppim de Oliveira

**Jose A. Puppim de Oliveira** is a faculty member at the Getulio Vargas Foundation (FGV/EAESP and FGV/EBAPE), and also teaches at the Federal University of Rio de Janeiro (COPPEAD-UFRJ), Fudan University (Shanghai) and Universidad Andina Simon Bolivar (UASB) in Quito, Ecuador. He is associated researcher at the United Nations University (UNU-IIGH) in Kuala Lumpur and the MIT Joint Program for Science and Policy for Global Change (USA).

## Reference

UN-Habitat (2011). *Cities and climate change: Global report on human settlements.* London/Washington, DC: Earthscan & UNCHS.

# Preface

Science is organized knowledge. Wisdom is organized life.

(Durant, 1938, p. 295–296)

Everyday problems can raise interesting questions for research. In 2011, when I was working in the urban planning division of an international consulting company, we were commissioned to prepare a state action plan for climate change for one of the Indian states. As spatial experts, we faced a peculiar challenge of suggesting climate strategies for cities and regions within a province without knowing their respective contributions to the problem (i.e. GHG responsibilities), and hence their appropriated actions in response to it. The absence of ready emission data and suitable methods for evaluation alluded reasoning that later inspired me to take such a practical problem into focused research at the doctoral level.

In fact, this query started gaining universal proportions and interest in a post-2009 scenario, when, for the first time, the world became more urban than rural in the history of humankind. As UN-Habitat (2011) famously put it, "climate change and urbanization are two of the greatest challenges currently facing humanity in the 21st century, whose effects are converging in dangerous ways". Back to back reports on cities and climate change by the three most influential international organizations – the World Bank, OECD and UN-Habitat – set the stage for a new policy discourse. They helped establish that cities contribute to climate change, and in turn get affected by it, and hence a comprehensive engagement is necessary in this interdisciplinary area. They also cautioned that as the world urbanizes and industrializes in the 21st century (the UN reports that 7 out of 10 people will live in cities by 2050), the situation would become even more alarming, particularly for developing countries.

Meanwhile, the national environment narrative in many of these developing countries, such as China, India, South Africa, Indonesia, etc., due to national development priorities and positioning in the international climate negotiations at the UNFCCC, was that of complete denial (i.e. we hardly contribute to the global carbon footprint). Similarly, for the states and cities in these countries, managing emissions was hardly a concern, much less a priority, in local planning and governance. It is this very complexity and contradictions in scientific and policy discourses in the global–local continuum that intrigued me to systematically and empirically investigate how carbon footprints behave while the world urbanizes, at different spatial units – global, national and local city level.

It is well acknowledged that GHGs of countries and cities have conventionally been studied as a consequence of their economic activity with 'income level' or GDP as the indicator, which merely represents one of the development indices and rather fuels debates on NS inequity. This research for the first time uses spatial analysis to add an additional dimension to this dynamic relation. This articulated into three basic research questions: (1) How do countries in different states of development – economic and urban/ rural – exhibit diverse patterns of GHG emissions? (2) What is the cumulative GHG responsibility or carbon footprint of urban areas within a country? (3) How do cities with different spatial dispositions affect their GHG emissions?

But the literature in this area highlighted more knowledge gaps than solutions, namely conceptual gap, empirical gap, methodological gap and policy-governance gap. Conceptual gap arises from the group of issues that inhibit a clear theoretical understanding of how nations, settlements or societies differently influence carbon footprint while they develop and urbanize. Empirical gap comes from either lack of sufficient scientific causations about urban carbon footprints or evidential data to support the prevailing theoretical knowledge. Methodological gap is caused by a gamut of reasons that inhibit proper assessment of a city's footprint. Policy-governance gap is a consequence of limited knowledge with cities about their roles and responsibilities to effectively respond and act for greener pathways (more on this, and how it helped frame five specific research questions and corresponding methods to attend to them, is discussed in Chapter 1). In fact, Chapters 2–6 systematically address each of the five research objectives.

The research is based on systematic understanding of several GHG methodologies, tools, protocols, etc. at multiple spatial scales, namely global, national urban–rural and city level. It involves intensive application of analytical and quantitative techniques such as spatial–numerical methods, spatial disaggregation technique, correlation and OLS regression, using multiple software such as Stata 2012, AutoCAD 2012, MS Excel, among others. In the process, this research generates a 3×3 spatial development matrix to analyse, for the first time ever, inequities in urban, economic, energy and GHG profiles of over 200 countries, superimposing data sets from the UN, World Bank, CDIAC and IEA. The results are confounding that, on the one hand, question the traditional duality between the Global North and South and, on the other, indicate sustained urban–rural disparity in addition to several other emerging patterns.

The national- and local-level findings are equally astounding, insightful and practically useful. They reason normative understanding with new and strong empirical evidence to inspire fresh ideas about urbanization and instruments that promote a low-carbon society and urban planning. This study attempts to answer, among others, the following questions: What is the most favourable rate of a country's urbanization? What should be the appropriate economies of scale for cities? What should be the most suitable spatial pattern for a city? What are appropriate regulatory, economic or governance mechanisms to achieve a low-carbon city?

The study is exemplary in multiple ways. There are limited studies across the globe that have tried to theoretically and empirically discern carbon throughputs on the urban–rural continuum, from the perspective of a physical or spatial planner. The challenge of this research could be gauged from the fact that this area of research is in its incipient stage in developing countries and has limited data sets for many indicators, particularly at the city level, in comparison to developed countries. The research attempts to methodologically tackle this by focused analysis at multiple spatial scales, namely the global, national urban–rural and city levels. The sampled cities have also been carefully selected to represent the diversity of location, geophysical conditions, city size, etc. within the urban–rural hierarchy.

The research contributes to the prevailing academic discourse and theories of urban metabolism, energy/ecological modelling, GHG accounting, spatial disaggregation, scale, economics, urban region, etc. At the same time, it feeds into practice with new methods/tools to energy and climate experts, economists and urban practitioners to rationally and equitably account for carbon throughputs of individual cities or cumulative urban areas within a country/region based on their spatial parameters. The book culminates with a series of suggestions that would facilitate economic policies, environmental strategies, spatial planning and an urban governance framework inclusive of green/low-carbon agenda.

The book is equally meant for thinkers and doers, those who prefer looking at the larger picture, value new knowledge for practical applications but leave no stone unturned when it comes to details. I have tried my best to make this volume as informative and simplified as possible. As mentioned earlier, the book is systematically structured such that each chapter (2–6) after the introduction responds to a definite research objective. All references are meticulously documented and provided consistently at the end of each chapter so that the reader does not have to struggle with an exhaustive bibliography at the end of the book. In addition, there are suggested readings for those who want to delve deeper into the subject. The book tests your learning after each chapter by asking basic and advanced sets of questions. This could prove helpful to graduate and postgraduate students preparing for examinations in this area. The most interesting and useful component is the *do it yourself exercises* designed at the end of almost every chapter. These should motivate readers to practice and master meticulous calculations in the simplest manner. While the book responds to several questions of the day, I hope it raises a spirit of wonder and interrogation among readers.

## Reference

Durant, W. (1938). *The Story of Philosophy* (new revised edition). New York: Garden City.

# Acknowledgements

This book is an embodiment of my doctoral research from 2011 to 2015. It is high time to acknowledge all those that have been instrumental in giving this its present shape and form. I am honoured and proud to be guided by individuals and institutions of different nationalities that deserve appreciation in equal measures and no particular order.

My heartiest thanks to Prof. Anjana Solanki and Dr. Vandana Sehgal. I am extremely grateful for their timely and persistent guidance, constructive criticism, thought-provoking discussions, encouragement and advice during this research. In the same tone, heartiest gratitude to Dr. Meenakshi Dhote, Professor, School of Planning & Architecture, New Delhi, who guided me in the earlier stages of this research. Early discussions with her were of immense value in scoping the study and framing research objectives. In this regard, I am also indebted to Prof. Subhakanta Mohapatra at Indira Gandhi National Open University, New Delhi, for his selfless guidance and co-authoring with me my first international publication in 2013. I hope my research association with him would have lasted more than that shaped by administrative constraints. I would like to offer my sincere gratitude to Prof O.P. Mathur, Senior Fellow, Institute of Social Sciences, and former Director, National Institute of Urban Affairs, for his support and motivation to research in a global and multidisciplinary context.

In May 2013, my research proposal earned me the coveted international PhD Fellowship in the Sustainable Urban Futures Programme at the United Nations University – Institute for the Advanced Study of Sustainability (UNU-IAS), Japan. I am grateful to my supervisor Dr. Jose A. Puppim de Oliveira (then Senior Research Fellow and Assistant Director at UNU-IAS and presently Visiting Research Fellow, UNU-IIGH, MSCP Scholar), whose genuine interest in climate change policy, environmental research and mentoring students has been a great inspiration. I would like to acknowledge his valuable suggestions in orienting my research with the universal discourse and constant motivation to present it at international forums. The formal seminars at UNU-IAS and that within our research group were instrumental in contextualizing my research arguments and present them to the global audience. Discussions in this area with international researchers such as Christopher Doll, Hooman Farzaneh, Csaba Pusztai, Magali Dreyfus and Aki Suwa used to be very constructive. I owe genuine regard for the technical, financial and academic support at UNU-IAS and its staff. I landed in Tokyo with my wife and 7-month-old daughter, and they made my research stay trouble-free. Ms. Taeko Morioka, Ms. Makiko, Ms. Akemi, Sayo, Mihoko, Bou Ty and several others were very helpful.

In spite of the hustle and bustle in its metropolis, Japan is known for its calm and serenity. The meditative and contemplative environs could soon become lonesome. But those were interesting times both inside and outside the Institute, as UNU-IAS and UNU-ISP were being merged into its present Institute for the Advanced Study of Sustainability. Indian researchers then formed the largest contingent at UNU-IAS, after the Japanese hosts. Prolonged interactions with Manu, Rama, Sohail, Yasmeen and Kabir over tea at odd hours filled the vacuum. My PhD fellowship period in Japan was one of the most learning, intensive and productive experiences, not only in my investigations on climate change and settlements, but more so in my life as well.

I am obliged to my present and former heads at the National Institute of Urban Affairs, Prof. Jagan A. Shah, Prof. Chetan Vaidya and Prof. Usha P. Raghupathi, for approving research activities and leaves, in spite of my official commitments. Special mention goes to Dr. Sabyasachi Tripathi, Research Fellow, and the economist who offered help with data analysis and interpretation while running the regression model. Some of the preliminary outcomes of the research were also sent for publication in peer-reviewed research journals, books, etc. Thanks is due to all anonymous reviewers whose comments and suggestions helped to augment and improve my research. I would like to acknowledge the support of my school, its staff and all those who have directly and indirectly contributed to this work. Early career researchers such as Neha G. Tripathi and Dipti Parashar deserve special mention for extending support whenever required.

The fruit of a work could not be appreciated in isolation from the context that yields it. Though my words would fell short in justification, I owe deepest gratitude to my parents for their extensive and unconditional support, patience and encouragement. In the same tone, warm thanks to my parents-in-law for their motivation to understand my research commitments over social obligations. Unfortunately, my father-in-law is not with us to see the fructification of this work, but I am sure we are blessed by him.

I owe exceptional appreciation to my wife, Shilpi, for her selfless support, physical and mental endurance (despite the health disorder developed during this while). With sheer wit and nerves, she stood beside me through thick and thin. As if the research was not engaging enough, this period saw the arrival of our two beautiful and adoring daughters – Prisha and Sanchi. I would like to give a toast to both of them, for their welcoming entry and beaming presence, showering moments of love, delight, appreciation, mischief and constant challenge. While Prisha was with us in Japan, for whom I used to keep waiting to sleep so I could burn the midnight oil, Sanchi is the new kid on the block.

Last but not the least, I give reverence to the innate force in me that drives me to work determinedly and try again after every failed attempt, without whom it would not have been possible to behold so much knowledge in such a small time and space. This phase of constant striving and uncertainty affirmed my faith. Thanks God, it's through!

# Abbreviations

| | |
|---|---|
| AAF | assumed activity factor |
| ADB | Asian Development Bank |
| ADEME | Agence d l'Environnement et de la Maîtrise de l'Energie |
| AFLOU | Agriculture, Forestry and Other Land Use |
| AFOLU | agriculture, forestry and other land use |
| AOSIS | Alliance of Small Island States |
| AR4 | Fourth Assessment Report (of the IPCC) |
| AR5 | Fifth Assessment Report (of the IPCC) |
| BBC | British Broadcasting Corporation |
| BC | black carbon |
| BS | Bharat Standard |
| C40 | Cities Climate Leadership Group |
| CAA | Constitutional Amendment Act |
| CAGR | Cumulative Aggregated Growth Rate |
| CBD | central business district |
| CBDR | common but differentiated responsibilities |
| CCP | Cities for Climate Protection |
| CDP | City Development Plan |
| CDIAC | Carbon Dioxide Information Analysis Center |
| CEA | Central Electricity Authority |
| CEMS | continuous emissions monitoring system |
| CFC | Central Finance Commission |
| CFCs | chlorofluorocarbons |
| $CH_4$ | methane |
| $CO_2$ | carbon dioxide |
| $CO_2e$ | $CO_2$ equivalents |
| COWI | Danish National Environmental Research Institute |
| CPCB | Central Pollution Control Board |
| ECBC | Energy Conservation Building Code |
| FAR | floor area ratio |
| F-gases | fluorinated gases |
| FOLU | forestry and other land use (FOLU) |
| FSI | Forest Survey of India |

| | |
|---|---|
| GCC | global climate change |
| GDP | gross domestic product |
| GEA | Global Energy Assessment |
| GHG | greenhouse gas |
| GIS | geographic information system |
| GoI | Government of India |
| GPC | Global Protocol for Community-Scale Greenhouse Gas Emissions |
| GPW | Gridded Population of the World |
| GRIHA | Green Rating for Integrated Habitat Assessment |
| GRIP | Greenhouse Gas Regional Inventory Protocol |
| GVA | gross value added |
| GWP | global warming potential |
| HEAT | Harmonized Emissions Analysis Tool |
| HFCs | hydrofluorocarbons |
| HPEC | High Powered Expert Committee on Urban Infrastructure |
| ICLEI | International Council for Local Environmental Initiatives |
| IEA | International Energy Agency |
| IEAP | International Local Government GHG Emissions Analysis Protocol |
| INCCA | Indian Network for Climate Change Assessment |
| INGO | international non-governmental organization |
| IO | input–output |
| IOA | input–output analysis |
| IPCC | Intergovernmental Panel on Climate Change |
| IPPU | industrial processes and product use |
| ISO | International Organization for Standardization |
| JNNSM | Jawaharlal Nehru National Solar Mission |
| JNNURM | Jawaharlal Nehru Urban Renewal Mission |
| KeTTHA | Kementerian Tenaga, Teknologi Hijau dan Air |
| LCA | Life Cycle Assessment |
| LDCs | Least Developed Countries |
| LDR | less developed regions |
| LULUCF | land use, land use change and forestry |
| MDR | more developed regions |
| MoEF | Ministry of Environment and Forests |
| MoUD | Ministry of Urban Development |
| MRV | Monitoring Reporting and Verification |
| $N_2O$ | nitrous oxide |
| NAPCC | National Action Plan on Climate Change |
| NATCOM | National Communication |
| NBC | National Building Code |
| NCEPC | National Committee on Environmental Planning and Coordination |
| NEP | National Environment Policy |

| | |
|---|---|
| NGO | non-governmental organization |
| NITI | National Institution for Transforming India |
| NIUA | National Institute of Urban Affairs |
| NMT | non-motorized transport |
| $NO_2$ | nitrogen dioxide |
| NS | North–South |
| NUTP | National Urban Transportation Policy |
| ODS | ozone-depleting substances |
| OECD | Organisation for Economic Co-operation and Development |
| OLS | ordinary least squares |
| PAT | Perform, Achieve and Trade |
| PEMS | predictive emissions monitoring system |
| PFCs | perfluorocarbons |
| PPP | public–private partnership |
| PUC | Pollution Under Control |
| RDF | refuse-derived fuel |
| REDD+ | Reducing Emissions from Deforestation and Degradation |
| SARC | Second Administrative Reforms Commission |
| SDGs | Sustainable Development Goals |
| $SF_6$ | sulfur hexafluoride |
| sq km | square kilometre |
| SRES | Special Report on Emissions Scenarios |
| TBIS | transboundary infrastructure supply chain |
| TCCCA | transparent, consistent, comparable, complete and accurate |
| TCPO | Town & Country Planning Organisation |
| TDR | Transferable Development Rights |
| TERI | The Energy and Resources Institute |
| UA | urban agglomeration |
| UAE | United Arab Emirates |
| UHI | urban heat islands |
| UK | United Kingdom |
| ULB | urban local body |
| UN | United Nations |
| UN DESA | United Nations Department of Economic and Social Affairs |
| UNDP | United Nations Development Programme |
| UNEP | United Nations Environment Programme |
| UNFCCC | United Nations Framework Convention on Climate Change |
| UNFPA | United Nations Population Fund |
| UN-Habitat | United Nations Human Settlements Programme |
| UR | urban–rural |
| US/USA | United States |
| USP | urban spatial parameters |
| USSR | Union of Soviet Socialist Republics |
| WBSCD | World Business Council for Sustainable Development |

| WEO | World Energy Outlook |
| WHO | World Health Organization |
| WMO | World Meteorological Organization |
| WRI | World Resources Institute |
| WTE | waste to energy |

# 1 Climate change and urban areas

> I saw an extraordinary panoply of objects flying past – bicycles, scooters, lamp posts, sheets of corrugated iron, even entire tea stalls. I buried my head in my arms and lay still . . . This was, in effect, the first tornado to hit Delhi – and indeed the entire region – in recorded meteorological history.
>
> (Ghosh, 2016)

Excerpts from Amitav Ghosh's new book, *The Great Derangement*, explain a sudden change of weather on the afternoon of 17 March 1978 in New Delhi, where the protagonist used to study at Delhi University while also working as a part-time journalist. The above extract is from an article published in the *Hindustan Times*, where the eminent author notes that nobody seems to be able to create a narrative around climate change. Indeed, the need to disseminate and involve publics is even more important now that current science has busted the myth around climate change.

## 1.1 Science of climate change

Climate change is an established scientific fact and a reality that has been witnessed constantly by mankind for the last few decades. The Intergovernmental Panel on Climate Change (IPCC) is the topmost scientific body that studies climate change at the global level. It asserts that warming of the climate system is unequivocal, and since the 1950s many of the observed changes are unprecedented over decades to millennia. The atmosphere and ocean have warmed, the amounts of snow and ice have diminished, and sea level has risen (IPCC, 2014). This is coupled with increased or irregular frequency, intensity and duration of weather events, causing floods, droughts and other storm surges, cyclones, etc. According to the IPCC:

> Climate change is a change in the state of the climate that can be identified (e.g., by using statistical tests) by changes in the mean and/or the variability of its properties, and that persists for an extended period, typically decades or longer.
>
> (IPCC, 2007b, p. 30)

## Box 1.1    Some extreme weather events and/or climate variability that have occurred across the world in recent memory

- UK 2000: Widespread flooding by wettest autumn since records began in 1766.
- Europe 2003: Record heatwave causes deaths of about 35,000 people.
- Middle East 2004: After unusually cold weather, snow falls in Dubai.
- India 2005: Deadliest floods in Maharashtra and Tamil Nadu claim over 5,000 lives.
- US 2005: Most active hurricane season on record. Katrina kills 1,300 in New Orleans.
- Brazil 2005: Worst drought in 60 years caused by lowest Amazon flow in 30 years.
- Canada 2005: Warmest summer on record.
- Horn of Africa 2006: Long-term drought and unprecedented floods in 50 years.
- India 2008: Heavy rainfalls trigger Kosi floods in Bihar.
- Argentina 2009: Exceptional heat sees record temperatures above 40° C.
- Russia 2010: Deadly heatwave sees wildfires and temperatures soaring in Moscow.
- Pakistan 2010: Worst floods in country's history affect 20 million people.
- China 2010: Torrential rains, some 1,500 people killed in one mudslide in north-west China.
- Australia 2010–2011: Worst floods in more than 50 years wreak havoc in north-east Australia.
- US 2011: Record number of wildfires, droughts and tornadoes kill hundreds of people.
- South America 2011: Landslides and floods kill hundreds in Brazil and Guatemala.
- China 2012: Temperatures in north-east China hit 43-year low at −15.3° C.
- Russia 2012: Floods kill 171 in south while west has hottest summer since 1500.
- Korea 2012: Both North and South Korea suffer their worst droughts on record.
- US 2012: Most extensive drought to affect the country since the 1930s.
- Europe 2012: Temperatures plummet to as low as −20° C in some cities.
- India 2013: Increasing trend of extremely heavy showers in rainfall data of last 104 years.
- Africa 2013: Heatwaves in Ghana and South Africa, and droughts in Namibia, Angola and Botswana.

- Australia 2013: Temperature rises up to 48° C, causing severe heatwave and bushfires.
- Asia 2013: Strongest typhoons such as Haiyan, Usagi and Phailin make landfall.
- Australia 2014: Warmest and driest ever weather in parts of Australia and New Zealand.
- Asia 2014: Cyclones Hudhud, Halong, Nakri and other torrential rains wreak havoc.
- America 2014: Extreme cold waves and heatwaves in Alaska, Canada, the US and Argentina.
- Europe 2014: Warmest year on record experienced across Europe.
- Arctic 2015: Heatwave in December spikes temperatures by 60° F above the norm.
- South America 2015: El Niño triggers floods in several countries, displacing 150,000 people.
- Ethiopia 2015: Prolonged droughts risk 10 million people in need of food aid.
- Yemen 2015: Over 1.1 million impacted and 40,000 displaced by rare cyclonic event.
- Asia 2015: Severe heatwave from Middle East to India kills thousands of people.

Munich Re (2011), the world's largest reinsurance company, has compiled global disaster records for 1980–2010. In its analysis, more than 90% of all disasters and 65% of associated economic damages were weather- and climate-related (i.e. high winds, flooding, heavy snowfall, heatwaves, droughts, wildfires). In all, 874 weather- and climate-related disasters resulted in 68,000 deaths and $99 billion in damages worldwide in 2010. The fact is that 2010 was one of the warmest years on record, as well as one of the most disastrous.

Climate change may be due to natural internal processes or *external forcings*, or due to persistent anthropogenic changes in the composition of the atmosphere or in land use. The IPCC claims that it is extremely unlikely (<5%) that the global pattern of warming during the past half-century can be explained without external forcing (i.e. it is inconsistent with being the result of internal variability), and very unlikely that it is due to known natural external causes alone. The apex international authority in climate governance, the United Nations Framework Convention on Climate Change (UNFCCC), in its Article 1, thus defines climate change as:

> A change of climate which is attributed directly or indirectly to human activity that alters the composition of the global atmosphere and which is in addition to natural climate variability observed over comparable time periods.
>
> (IPCC, 2007c, p. 996)

The distinct feature of the UNFCCC definition is that its area of concern or "attribution to climate change" is particularly focused on additionalities to natural climate variability, caused by man-made factors. It affirms that climate change is caused by the sustained presence of greenhouse gases (GHGs) in the atmosphere. The IPCC defines GHGs as:

> those gaseous constituents of the atmosphere, both natural and anthropogenic, that absorb and emit radiation at specific wavelengths within the spectrum of thermal infrared radiation emitted by the Earth's surface, the atmosphere itself, and by clouds. This property causes the greenhouse effect.
>
> (IPCC, 2007b, p. 82)

A GHG assessment, also called an inventory, focuses on anthropogenic (i.e. man-made) emissions. Water vapour, for instance, is the gas that has the greatest impact on the greenhouse effect. However, the atmospheric water vapour concentration is not substantially affected by human activities, thus water vapour is commonly not referred to as a major anthropogenic greenhouse gas. Some greenhouse gases in the atmosphere are entirely man-made, such as the halocarbons and other chlorine- and bromine-containing substances. The Montreal Protocol deals with these gases. The Kyoto Protocol refers to the following gases: $CO_2$, $N_2O$, $CH_4$, $SF_6$, hydrofluorocarbons (HFCs) and perfluorocarbons (PFCs). These six 'Kyoto gases' are supposed to be the most important anthropogenic gases with regard to the greenhouse effect (IPCC, 2007b).

It is a known fact that industrialization in the last 200 years has led to mass consumption and burning of fossil fuels for industrial production, metallurgical operations, cement production, rail and road transportation, thermal power generation and allied activities, which has been a key trigger for rising local pollution in the ambient air and GHGs in the upper layers of the atmosphere. Data analysis from 1751 to 2012 by Earth Policy Institute (2013) shows that over the last century, emissions from fossil fuel burning, which forms the biggest bunch of the GHG basket, have mounted sevenfold from about 1,000 $MtCO_2e$ in the 1920s to over 6,500 $MtCO_2e$ in 2000. In order to evaluate the actual or estimated impacts of GHG emissions from a particular human being, anthropogenic or socio-economic activity, industrial process, etc., the concept of carbon footprint is increasingly being utilized:

> Carbon footprint of a functional unit is the climate impact under a specified metric that considers all relevant emission sources, sinks, and storage in both consumption and production within the specified spatial and temporal system boundary.
>
> (Peters, 2010, p. 245)

The IPCC defines carbon footprint as "The total set of greenhouse gas emissions caused by an organization, event, product or person" (IPCC, 2007b). It is expressed in carbon dioxide, or its equivalent of other GHGs emitted. Sources of carbon footprint are invariably the same, responsible for GHG emissions (i.e. transport, land clearance, production and consumption of food, fuels, manu-factured goods, materials, wood, roads, buildings and services). The concept of carbon footprint actually originates from the discourse on ecological footprint and Life Cycle Assessment (LCA) of natural resources. Ecological footprint is meas-ured as a total of six factors: cropland footprint, grazing footprint, forest footprint, fishing ground footprint, carbon footprint and built-up land, and expressed as the amount of productive land and sea area required to sequester carbon emissions on a per capita basis (Rees, 1992). It is represented by the following basic equation:

$$\text{ecological footprint } (X) = \frac{\text{global hectares affected by humans}}{\text{population}}$$

So, is carbon emissions (footprint) different from GHG emissions (footprint)? This could be better understood from the concept of $CO_2e$ and GWP (Box 1.2).

---

## Box 1.2 Concept of $CO_2e$ and GWP

In view of making the climate impact of different GHGs comparable, they are normally converted to $CO_2$ equivalents ($CO_2e$). $CO_2$ is thereby the reference gas against which other gases are measured, and has a global warming potential (GWP) of 1. The global warming potential represents how much a certain mass of a gas contributes to global warming compared to the same mass of $CO_2$. It is based on the different times "gases remain in the atmosphere and their relative effectiveness in absorbing outgoing ther-mal infrared radiation" (IPCC, 2007b, p. 81). For instance, nitrous oxide is 310 times more potent than $CO_2$. A ton of nitrous oxide can thus be converted to $CO_2$ equivalents by multiplying it by 310. Inventories that cover different GHGs commonly display results in $CO_2$ equivalents. In this respect, it is of crucial importance that the sources and values on which the calculation of these equivalents is based are made transparent. This can be illustrated by the example of the time horizon used for the calculation of the global warming potential. Some gases remain only for short periods of time in the atmosphere, whereas other gases can remain for thousands of years. Thus, different time horizons lead to different global warming potentials. Methane, for instance, has, on average, a shorter lifetime in the atmosphere than $CO_2$. If the calculation of the global warming potential of methane is based on a time horizon of 20 years, methane has a global warming potential of 72 (i.e. 72 times greater than $CO_2$). A time horizon of 100 years yields a global warming potential of 25, and a time horizon of 500 years yields a global warming potential of 7.6 (IPCC, 2007c). National inventories that are compiled according to the IPCC guidelines use a time horizon of 100 years.

## 1.2 Evolution of climate change research

Climate change has been observed persistently by scientists since the last century, but post-1988 has seen unprecedented global efforts aimed towards mitigation and adaption. There is growing research in not just observing, monitoring and predicting weather patterns, extreme events and their risks, vulnerabilities and impacts on natural and human environments, economic sectors, etc, but also on multilevel and multisectoral interventions needful to deal with the impacts. Evolution of knowledge in climate change, both in the science and policy realms, in the recent past is presented in Table 1.1.

*Table 1.1* Evolution of climate change research – science and policy timeline

| Year | Science/ policy | Event |
|------|-----------------|-------|
| 1972 | Policy | Setting up of National Committee on Environmental Planning and Coordination (NCEPC). |
| 1976 | Science | Studies reveal CFCs (1975), methane and ozone (1976) seriously contribute to the greenhouse effect. |
| 1981 | Policy | Enactment of Air (Prevention and Control of Pollution) Act. |
| 1986 | Policy | Enactment of Environmental Protection Act. |
| 1987 | Policy | Montreal Protocol of the Vienna Convention imposes international restrictions on emission of ODS. |
| 1988 | Science/ policy | WMO and UNEP establish the Intergovernmental Panel on Climate Change (IPCC). |
| 1998 | Policy | Enactment of Motor Vehicles Act. |
| 1992 | Policy | Conference in Rio de Janeiro produces UN Framework Convention on Climate Change, but the US blocks it and calls for serious action. |
| 1995 | Science/ policy | IPCC 2nd Assessment Report detects 'signature' of human-caused greenhouse effect, declares serious global warming likely in coming century. |
| 1998 | Science | India's inventory for 1990 level emissions using 1995/96 IPCC Guidelines for national inventories. |
| 2001 | Science/ policy | IPCC 3rd Assessment Report finds "new and stronger evidence" that humanity's GHG emissions are main cause of warming seen in the second half of the 20th century. |
| 2002 | Policy | India ratifies the Kyoto Protocol. |
| 2003 | Policy | Enactment of The Electricity Act. |
| 2004 | Science | India releases NATCOM-I (1st GHG inventory) to the UNFCCC. |
| 2004 | Policy | Pollution Under Control (PUC) norms are initiated for motor vehicles. |
| 2005 | Policy | Kyoto Treaty goes into effect, signed by major industrial nations except the US, Japan and Western Europe. $CO_2$ levels in atmosphere reach 380 ppm. |
| 2005 | Policy | Release of National Environment Policy, National Electricity Policy, National Urban Transport Policy. |
| 2007 | Science/ policy | IPCC 4th Assessment Report concludes more than 90% likely that humanity's emissions of GHG are responsible for modern-day climate change. |
| 2007 | Policy | Establishment of Energy Conservation Building Code (ECBC). |
| 2008 | Policy | India launches National Action Plan on Climate Change. |

| 2009 | Policy | China overtakes the US as the world's biggest greenhouse gas emitter, although the US remains well ahead on a per capita basis. |
|------|--------|---|
| 2010 | Policy | Launch of Jawaharlal Nehru National Solar Mission (JNNSM). |
| 2010 | Science | India releases NATCOM-II (2nd GHG inventory) to the UNFCCC. |
| 2012 | Policy | Launch of India's pilot on UN mechanism for REDD+ (Reducing Emissions from Deforestation and Degradation) and Perform, Achieve and Trade (PAT) mechanism. |
| 2014 | Science/ policy | IPCC 5th Assessment Report released. |
| 2014 | Policy | Over 200 countries convene for UNFCCC in Bonn (4–15 June). |
| 2015 | Policy | UNFCCC decides to adopt Paris Agreement at Conference of Parties 21. |
| 2016 | Policy | Sustainable Development Goals (SDGs) adopted as global agenda for 2030. Goal 11 and Goal 13 of SDGs particularly focus on Cities and Climate Change, respectively. |

Source: BBC, WRI, UNEP and other websites

## 1.3 Causes of climate change: gases and sectors

The causation of climate change is attributed to growing emissions of different gases (see Figure 1.1c), empirically assessed from different economic sectors. At the global scale, key GHGs emitted by human activities are (IPCC, 2014):

- **Carbon dioxide ($CO_2$):** Fossil fuel use is the primary source of $CO_2$. The way in which people use land is also an important source of $CO_2$, especially when it involves deforestation. Land can also remove $CO_2$ from the atmosphere through reforestation, improvement of soils, and other activities. $CO_2$ emissions from fossil fuels and industrial processes contributed to 65% of GHG emissions in 2010, while forestry and other land use (FOLU) accounted for another 11%.
- **Methane ($CH_4$):** Agricultural activities, waste management and energy use all contribute to $CH_4$ emissions. It had a share of 16% of GHG emissions in 2010.
- **Nitrous oxide ($N_2O$):** Agricultural activities, such as fertilizer use, are the primary source of $N_2O$ emissions. It contributed to 6.2% of GHG emissions in 2010.
- **Fluorinated gases (F-gases):** Industrial processes, refrigeration and the use of a variety of consumer products contribute to emissions of F-gases, which include hydrofluorocarbons (HFCs), perfluorocarbons (PFCs) and sulfur hexafluoride ($SF_6$). Collectively, the share of fluorinated gases was 2% of GHG emissions in 2010.

Black carbon (BC) is a solid particle or aerosol, not a gas, but it is also known to contribute towards the warming of the atmosphere.

The main sectors for which emissions can be assessed – energy for electricity, transportation, commercial and residential buildings, industry, waste, agriculture, land use change and forestry – are all relevant to urban areas, which rely on

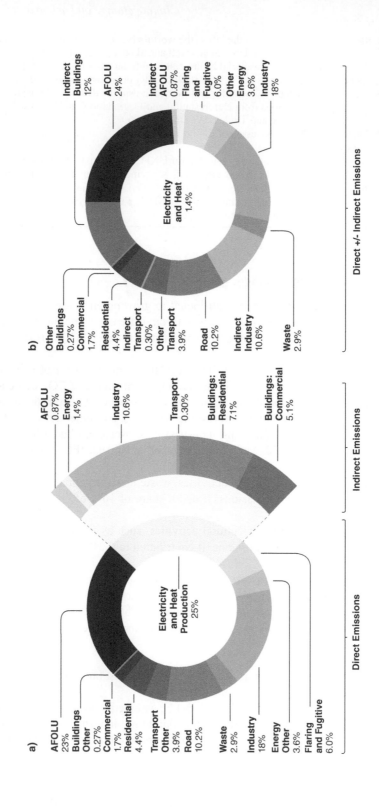

a)

**Direct Emissions**

AFOLU 23%
Buildings Other 0.27%
Commercial 1.7%
Residential 4.4%
Transport Other 3.9%
Road 10.2%
Waste 2.9%
Industry 18%
Energy Other 3.6%
Flaring and Fugitive 6.0%

Electricity and Heat Production 25%

**Indirect Emissions**

AFOLU 0.87%
Energy 1.4%
Industry 10.6%
Transport 0.30%
Buildings: Residential 7.1%
Buildings: Commercial 5.1%

b)

Other Buildings 0.27%
Commercial 1.7%
Residential 4.4%
Indirect Transport 0.30%
Other Transport 3.9%
Road 10.2%
Indirect Industry 10.6%
Waste 2.9%

Electricity and Heat 1.4%

Indirect Buildings 12%
AFOLU 24%
Indirect AFOLU 0.87%
Flaring and Fugitive 6.0%
Other Energy 3.6%
Industry 18%

**Direct +/- Indirect Emissions**

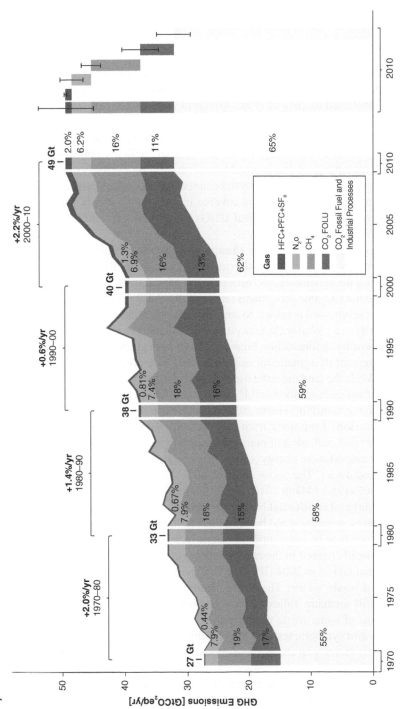

c)

*Figure 1.1* (a) Allocation of total GHG emissions in 2010 (49.5 GtCO₂eq/yr) across the five sectors. Pull-out from panel A allocates indirect CO₂ emission shares from electricity and heat production to the sectors of final energy use. (b) Allocates that same total emissions (49.5 GtCO₂eq/yr) to reveal how each sector's total increases or decreases when adjusted for indirect emissions. (c) Total annual GHG emissions by groups of gases 1970–2010, along with estimated uncertainties illustrated for 2010 (whiskers). The uncertainty ranges provided by the whiskers for 2010 are illustrative given the limited literature on GHG emission uncertainties

Source: a) and b) IPCC (2014), c) Victor et al. (2014)

goods, services and processes taking place both inside and outside their boundaries (UN-Habitat, 2011, p. 61). Accordingly, global greenhouse gas emissions can also be broken down by the economic activities that lead to their production, as follows (see Figures 1.1a and 1.1b):

- **Energy supply:** The burning of coal, natural gas and oil for electricity and heat is the largest single source of global greenhouse gas emissions. The sector contributed to 26% of global GHGs in 2004 (IPCC, 2007a) and 25% in 2010 (IPCC, 2014).
- **Industry:** Emissions from industry primarily involve fossil fuels burned on-site at facilities for energy. This sector also includes emissions from chemical, metallurgical and mineral transformation processes not associated with energy consumption. It needs to be noted that emissions from electricity use are excluded, and are instead covered in the energy supply sector. The sector accounted for 19% of global GHGs in 2004 (IPCC, 2007a) and 21% in 2010 (IPCC, 2014).
- **Agriculture, land use, land use change and forestry:** Emissions from this sector primarily include carbon dioxide ($CO_2$) emissions from deforestation, land clearing for agriculture, and fires or decay of peat soils. This estimate does not include the $CO_2$ that ecosystems remove from the atmosphere. The amount of $CO_2$ that is removed is subject to large uncertainty, although recent estimates indicate that, on a global scale, ecosystems on land remove about twice as much $CO_2$ as is lost by deforestation. Emissions from agriculture mostly come from the management of agricultural soils, livestock, rice production and biomass burning. While the land use, land use change and forestry sector share was 17% and that of agriculture stood at 14% of global GHGs in 2004 (IPCC, 2007a), their collective contribution was estimated at 24% in 2010 (IPCC, 2014).
- **Transportation:** Emissions from this sector primarily involve fossil fuels burned for road, rail, air and marine transportation. Almost all (95%) of the world's transportation energy comes from petroleum-based fuels, largely gasoline and diesel. The sector contributed to 13% of global GHGs in 2004 (IPCC, 2007a) and 14% in 2010 (IPCC, 2014).
- **Commercial and residential buildings:** Emissions from this sector arise from on-site energy generation and burning fuels for heat in buildings or cooking in homes. It needs to be noted that emissions from electricity use are excluded, and are instead covered in the energy supply sector. The sector accounted for 8% of global GHGs in 2004 (IPCC, 2007a) and 6.4% in 2010 (IPCC, 2014).
- **Waste and waste water:** The largest source of GHG emissions in this sector is landfill methane, followed by waste water methane and nitrous oxide. Incineration of some waste products that were made with fossil fuels, such as plastics and synthetic textiles, also results in minor emissions of $CO_2$. The share of this sector was about 3% of global GHGs in 2004 (IPCC, 2007a).

The course by which different economic or development sectors render various activites/end use, and in the process emit different gases, is evident in Figure 1.2.

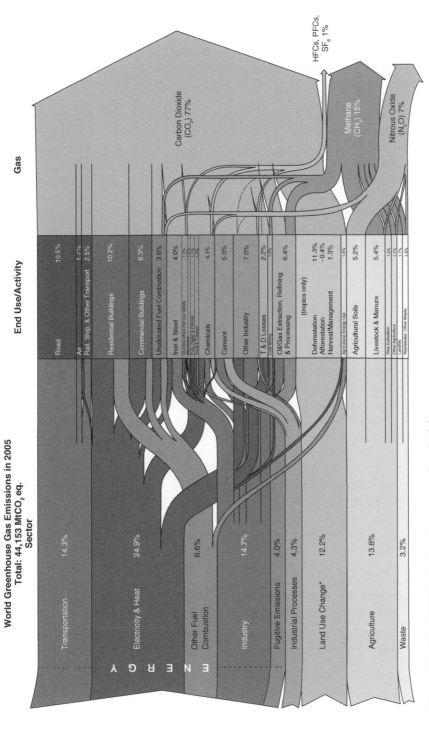

*Figure 1.2* Global GHG emissions by sector, end use/activity

Source: World Resources Institute (2016)

## 1.4 Spatial perspective of climate change: global pattern

Though climate change is a global phenomenon, scientific research informs that its impacts are neither equal nor consistent everywhere on the Earth. The IPCC reports spatial disparity in global environmental changes – intensity, frequency and regional distribution of impacts, as evident from the changes observed in the world's average surface temperature and average precipitation from 1986–2005 to 2081–2100 (see Figure 1.3).

Likewise, the cause of climate change rooted in excessive emission of GHGs to the atmosphere is also not equal or consistent over the space. Certain countries, regions and societies, owing to higher levels of industrialization, development, fossil fuel use, consumption of goods and services, etc., contribute more GHGs than others (the relationship between state of development, urbanization and GHG emissions is further explored in Chapter 2). While 18% of the world's population living in developed countries account for 47% of global $CO_2$ emissions, 82% of the world's population living in developing countries account for the remaining 53% (Rogner *et al.*, 2007). The inequity runs deep in the global climate regime of the Kyoto Protocol, as evident in the distribution of regional per capita GHG emissions, according to the population of Annex I countries (essentially the developed world) and Non-Annex I countries (the developing world). The AR4 Synthesis Report reveals that though Annex I countries constitute 19.7% of the world population, their average GHG emission stands tall at 16.1 t $CO_2$e/capita in 2004. Conversely, Non-Annex I countries constitute 80.3% of the world population but have average GHG emissions of 4.2 t $CO_2$e/capita (IPCC, 2007a). This demonstrates that, like the impacts of climate change, the GHG contribution or carbon footprint of nations also varies significantly over space.

Even greater differences can be seen if individual countries are compared: per capita $CO_2$e emissions vary from less than 1 ton (t) per year for Bangladesh and Burkina Faso to more than 12 t for Canada, the US and Australia and above 50 t for Qatar. The carbon footprint of different countries, compiled by the World Resources Institute from a variety of sources, including data from the Carbon Dioxide Information Analysis Center (CDIAC), is shown in Figure 1.4. It reveals that North America, Europe and Asia have the biggest carbon/GHG footprints.

There is a sufficient degree of information and debate on the historical trends of these GHGs. The literature steers the thought of fairness, equity and justice in appropriation of global environmental burdens. As a consequence of the Industrial Revolution, the world has witnessed huge amounts of carbon emissions from the early industrialized countries such as the United States (US), Germany, the former USSR, the United Kingdom (UK), Japan, France, Canada, Italy, Mexico and Australia, and followed recently by emerging economies such as China, India, Brazil, South Africa and Indonesia. Coincidentally, the assessment reports of the IPCC mandated by the UNFCCC to steer climate change research and negotiation at the global level steer clear of the political and historic overtones, and are devoid of any particular references to this, though a lot of international climate discourse, scientific publications and popular literature accessible on the

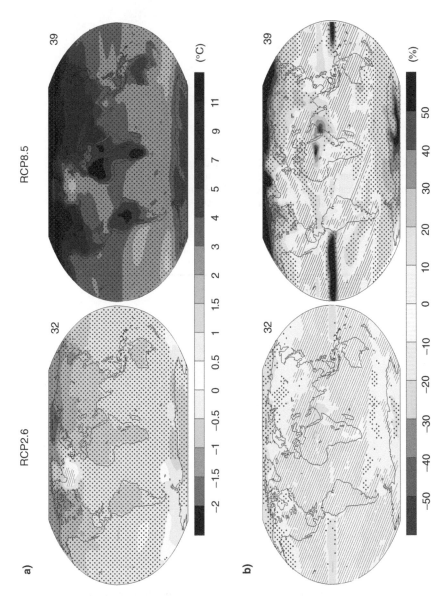

*Figure 1.3* (a) Change in average surface temperature 1986–2005 to 2081–2100. (b) Change in average precipitation 1986–2005 to 2081–2100

Source: IPCC (2014)

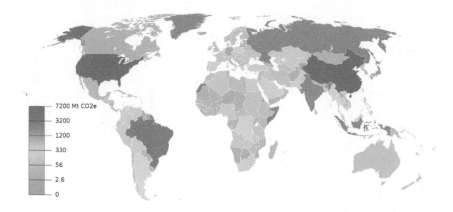

*Figure 1.4* GHG emissions by country in 2005, including land use change (based on data for carbon dioxide, methane, nitrous oxide, perfluorocarbon, hydrofluorocarbon and sulfur hexafluoride emissions compiled by the WRI from a variety of sources, including CDIAC)

Source: World Resources Institute (2016)

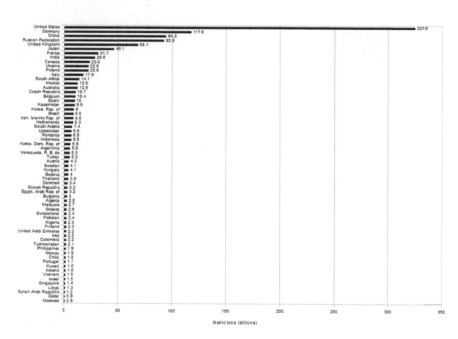

*Figure 1.5* Cumulative energy-related carbon dioxide emissions between 1850 and 2005 (in t/capita)

Source: Enescot (2011)

Internet speaks volumes about it. Figure 1.5 is one such diagram that illustrates the variations in cumulative energy-related carbon dioxide emissions of different countries (expressed in t/capita) between 1850 and 2005.

As evident, there is a strong historical, political, economic and spatial perspective to the carbon footprint that has traditionally defined the North–South (NS) divide in international climate discourse. This fuels the debate and interest in allocation of responsibility for global carbon emissions. The discourse churning international policy is guided by the principle of "common but differentiated responsibilities" (CBDR) enshrined within the Kyoto Protocol and the UNFCCC 1992. As such, there is enormous spatial inequity evident at the international level, where individual countries represent aggregated data on GHG emissions. Considering a similar though greater degree of diversity exists in geographical locations, economic activities, fossil fuel, technologies and consumption patterns of local populations within the same country, an analogous argument for CBDR at the subnational or settlement level is palpable. There is bound to be an even greater disparity and diversity in carbon footprints of urban and rural areas at the subnational level across the globe.

## 1.5 Role of urban areas in contributing to climate change

The link between climate change and human settlements has already been established. Cities contribute to climate change and cities are affected by climate change (UN-Habitat, 2010). It has been recognized that climate change and urbanization are two of the greatest challenges currently facing humanity in the 21st century, whose effects are converging in dangerous ways. Another UN study claims that cities are responsible for 75% of global energy consumption and 80% of GHGs (UN, 2007). The World Bank report entitled *Cities and Climate Change: An Urgent Agenda* further asserts that cities are major contributors to greenhouse gas emissions, so much so that the 50 largest world cities combined rank 3rd in both population and GHG emissions, and 2nd in GDP when compared to the largest and wealthiest countries. Energy-wise, the world is already predominantly urban. Cities meet approximately 72% of their total energy demand from coal, oil and natural gas – the main contributors to GHGs (World Bank, 2010). The International Energy Agency (IEA), in its report Global Energy Assessment (GEA, 2012) estimates that between 60 and 80% of final energy use globally is urban, with a central estimate of 75%. There is a minor difference in the reporting of results between the two studies, but undoubtedly there is a clear consensus on the degree of urban responsibilities or burdens at the global level.

The literature underpins the importance of considering the contribution of urban areas to climate change, for several reasons (UN-Habitat, 2011). First, a range of activities is associated with cities and their functioning that contribute to GHG emissions. Transportation, energy generation and industrial production within the territorial boundaries of towns and cities generate GHG emissions directly. Urban centres rely on inward flows of food, water and consumer goods and services that may result in GHG emissions from areas outside the city. Second, measuring emissions from different cities provides a basis for comparisons to be made and the potential for interurban competition and cooperation. Climate-friendly development has the potential to attract external investment, and the growing importance of international urban networks provides spaces for learning and knowledge sharing. Emission measuring has recently been inserted into global policy debates. For example, the United Nations Environment Programme (UNEP), the United Nations Human Settlements Programme (UN-Habitat) and the World Bank launched an International Standard for Determining Greenhouse Gas Emissions for Cities at the World Urban Forum in Rio de Janeiro in March 2010. Third, an assessment of the contribution of cities to climate change is a vital first step in identifying potential solutions. The large and growing proportion of the Earth's population living in towns and cities, and the concentration of economic and industrial activities in these areas, means that they need to be at the forefront of mitigation. The establishment of emission baselines is necessary if effective mitigation benefits are to be identified and applied technologically, and, last but not the least, internalizing it with local development priorities, projects and urban governance framework.

---

**Box 1.3**

**(a)    Urban concepts and definitions**

**Urban:** Settlements or localities defined as 'urban' by national statistical agencies.

**Urbanization:** The process of transition from a rural to a more urban society. Statistically, urbanization reflects an increasing proportion of the population living in settlements defined as urban, primarily through net rural-to-urban migration. The *level of urbanization* is the percentage of the total population living in towns and cities, while the *rate of urbanization* is the rate at which it grows (UNFPA, 2007).

**Urban growth:** The increase in the number of people who live in towns and cities, measured either in relative or absolute terms (UNFPA, 2007).

**Megacity:** UN-Habitat defines a megacity as a city with 10 million or more inhabitants.

**(b)    Why a spatial perspective is important**

Spatial analysis or spatial statistics includes any of the formal techniques that study entities using their topological, geometric or geographic properties. The phrase properly refers to a variety of techniques, many still in their early development (http://en.wikipedia.org/wiki/Spatial_analysis,

---

accessed 30 December 2015). Geographic information systems (GISs) and the underlying geographic information science that advances these technologies have a strong influence on spatial analysis. Various aspects of spatial studies include the following:

**Spatial location:** Transfer positioning information of space objects with the help of a space coordinate system. Projection transformation theory is the foundation of spatial object representation.

**Spatial distribution:** Similar spatial object groups positioning information, including distribution, trends, contrast, etc.

**Spatial form:** Geometric shape of the spatial objects.

**Spatial space:** Space objects' approaching degree.

**Spatial relationship:** Relationship between spatial objects, including topological, orientation, similarity (such as land use), etc.

On a temporal scale, growing GHG emissions show a strong association with the urbanization process too. Just 200 years ago, a mere 3% of the Earth's inhabitants lived in cities, but around the year 2009, for the first time in history, city dwellers outnumbered rural inhabitants (UN DESA, 2011). The climate future under the present scenarios seems challenging, as exploding urban growth is imminent. People living in cities across the globe are expected to double by 2050, while the built-up area is expected to triple during the same period (World Bank, 2010). According to UN estimates, 6.4 billion people living in cities by

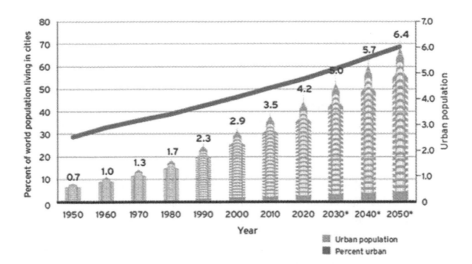

*Figure 1.6* People living in cities (percentage of world population and total)

Source: World Bank (2010)

2050 will constitute 70% of the population (UN DESA, 2012). All of the above forms an essential base – urban planning in the next few decades will determine the pace of global warming.

At the same time, it has been argued that cities also offer a solution to the problem of GHGs in several ways. First, urban authorities and local governments have the potential to implement mitigation programmes effectively, because of the type of responsibilities they hold in relation to land use planning, local public transportation and the enforcement of industrial regulations (Dodman, 2009a). Urban authorities the world over in the recent past have set ambitious targets for emissions reductions, such as London (UK), Tokyo (Japan), Copenhagen (Denmark) and Toronto (Canada), to name a few. Second, the concentration of people and industries in large cities provides the opportunity for technological innovations, such as combined heat and power and waste-to-energy generation plants that can generate electricity more efficiently, and it also makes mass transit systems more cost- and time-effective. Third, this concentration also provides the opportunity for the rapid spread and adoption of new ideas and innovations, both in technical and behavioural solutions. Finally, reducing greenhouse gas emissions can also lead to various other local benefits. The author illustrates it with examples of reducing fuel costs, better pedestrian safety and improved public health through reduced air pollution.

The Charter of European Cities and Towns Towards Sustainability (as approved by participants at the European Conference on Sustainable Cities and Towns in Aalborg, Denmark, on 27 May 1994) accepts the Responsibility for the Global Climate (para 1.10). It states:

> We, cities and towns, understand that the significant risks posed by global warming to the natural and built environments and to future human generations require a response sufficient to stabilize and then to reduce emissions of greenhouse gases into the atmosphere as soon as possible. It is equally important to protect global biomass resources, such as forests and phytoplankton, which play an essential role in the Earth's carbon cycle. The abatement of fossil fuel emissions will require policies and initiatives based on a thorough understanding of the alternatives and of the urban environment as an energy system. The only sustainable alternatives are renewable energy sources.
>
> (Aalborg, 1994)

The recent global development paradigm duly recognizes the growing importance of urban settlements in addressing the planetary challenge of global warming and climate change. The Sustainable Development Goals (SDGs), which build upon the erstwhile Millennium Development Goals, follow a multipronged approach to achieve development of all humankind within the planetary boundaries. They bear some new themes, such as affordable energy, sustainable economic growth, industrialization, climate change and its impacts, marine resources, terrestrial ecosystems, inclusive and just societies, global partnerships, and aims to make cities inclusive, safe, resilient and sustainable (Goal 11), commonly known as the *Cities Goal* or the *Urban SDGs*. In addition, SDGs also target to "ensure sustainable consumption and production patterns" (Goal 12) and to "take urgent action to combat climate change and its impacts" (Goal 13). The inclusion of the urban theme in the global development narrative affirms the message of Dr. Joan Clos, Executive Director of UN-Habitat: "Our struggle for global sustainability will be won or lost in cities". It could be reasoned with credible evidence in support, as across the globe cities and towns of various size and function create an unprecedented demand for fossil carbon to fuel their national economies.

---

**Box 1.4  Targets of Goal 11: to make cities inclusive, safe, resilient and sustainable**

- By 2030, ensure access for all to adequate, safe and affordable housing and basic services and upgrade slums.
- By 2030, provide access to safe, affordable, accessible and sustainable transport systems for all, improving road safety, notably by expanding public transport, with special attention to the needs of those in vulnerable situations, women, children, persons with disabilities and older persons.
- By 2030, enhance inclusive and sustainable urbanization and capacity for participatory, integrated and sustainable human settlement planning and management in all countries.
- Strengthen efforts to protect and safeguard the world's cultural and natural heritage.
- By 2030, significantly reduce the number of deaths and the number of people affected and substantially decrease the direct economic losses relative to global gross domestic product caused by disasters, including water-related disasters, with a focus on protecting the poor and people in vulnerable situations.
- By 2030, reduce the adverse per capita environmental impact of cities, including by paying special attention to air quality and municipal and other waste management.
- By 2030, provide universal access to safe, inclusive and accessible, green and public spaces, in particular for women and children, older persons and persons with disabilities.
- Support positive economic, social and environmental links between urban, peri-urban and rural areas by strengthening national and regional development planning.
- By 2020, substantially increase the number of cities and human settlements adopting and implementing integrated policies and plans towards inclusion, resource efficiency, mitigation and adaptation to climate

*(continued)*

*(continued)*

>    change, resilience to disasters, and develop and implement, in line with
>    the Sendai Framework for Disaster Risk Reduction 2015–2030, holistic
>    disaster risk management at all levels.
>
> •  Support least developed countries, including through financial and
>    technical assistance, in building sustainable and resilient buildings uti-
>    lizing local materials.
>
> Source: www.globalgoals.org/global-goals/sustainable-cities-and-communities/
> (accessed 11 February 2017)

Collectively, the above three goals tend to attain sustainable practices in our industrial processes, business activities and urban lifestyles, thereby reducing the carbon footprint. But the application and interpretation of the SDGs in countries and regions would vary according to their local circumstances. It is this hierarchical or local relevance of global principles that any research in this multidisciplinary area needs to reflect upon. For instance, in context of India, empirical and policy studies on cities and climate change are few and far between. One such research (Revi, 2008) proposes a climate agenda for cities in India. It considers the likely changes that climate change will bring in temperature, precipitation and extreme rainfall, drought, river and inland flooding, storms/storm surges/coastal flooding, sea level rise and environmental health risks, and who within urban populations are most at risk. It highlights the importance of today's infrastructure investments, taking into account climate change, given the long lifespan of most infrastructure, and the importance of urban management engaging with changing risk profiles. One of the first and most pertaining scientific studies associating climate change with Indian cities is highly adaptation-centric. It culminates by describing a possible urban climate change adaptation framework, including changes needed at the national, state, city and neighbourhood levels, and linkages to mitigation. Nevertheless, a similar but comprehensive mitigation-centric climate change framework for Indian cities is highly needful.

## 1.6 Scope of study: the case of India

As discussed above, climate change study may broadly be classified into adaptation and mitigation. This research is broadly aimed at deciphering the mitigation aspects of climate change. The case study is that of one of the most rapidly urbanizing countries in the developing world, India. One might be interested to reason how a comprehensive study of Indian cities is imperative to its climate change narrative. In fact, there are several explanations to this query. First, India is one of the largest contributors to GHGs that cause global warming after China ($11,182$ $MtCO_2e$) and the US ($6,715$ $MtCO_2e$) (UNEP, 2012). It has become a time-honoured understanding among the international community that though the cause and implication of climate change are global, local action for mitigation and adaptation is paramount (OECD, 2009; UN-Habitat, 2011). The worldwide emergence of decentralization and local governments in the last two decades is at the core of this phenomenon. Hence, the role of cities in their contribution towards GHG emissions is gaining consciousness, as they are traditionally considered to be engines of industrial growth, and post-2009 resumed significance when, for the first time in the history of human civilization, urban dwellers outnumbered their rural counterparts (UN DESA, 2011). In order to appreciate the need to study Indian cities' emissions, it is vital to comprehend India's growing carbon throughput against the world while it is urbanizing. The major observations, inter alia, include high rates of urbanization considered as an outcome of industrialization (akin to climate change), known to be particularly affecting low-income and middle-income nations, such as India.

Second, the Indian case appropriately represents the situation of many developing countries (ie. it is on an economic development pathway, facing the dilemma of conservation and development, and has a multi-governance set-up to deal with). Being about 30% urbanized, there is a general notion that urban contributions are insignificant. On the other hand, there is an immense potential to capture future locked-in emissions in building, transport and energy sectors.

Third, the relevance of the study is rooted in the existing political and economic realities. While the Constitution of India declares the nation to be, among other virtues, a socialist, democratic republic, as per recent data on development indices, 20.6% of the population lives below the poverty line (Chandy and Kharas, 2014). At the same time, approximately 169 million were urban poor in 2005, which is projected to increase to 202 million by 2020 (UN-Habitat, 2006). Traditionally, the state's urban policies have focused on demand-centric issues of basic services such as water, sanitation and shelter. Climate change strategy, on the other hand, is a good-good soft policy that usually gets administered from top-down dictates such as international commitments, national environmental policy, etc. Abating climate change and reducing GHGs at the ground level is hence hardly an agenda, and even less a priority, for local governments, state ministries, line departments, etc. Such state-induced complacency and indifference pose a high risk of inaction. As such, it becomes even more pertinent to scope an inquiry in this grey area. This forms a basis for this research on India.

## 1.7 Prevailing knowledge gaps

It is acknowledged that urban areas – which are now home to more than half of the world's population – clearly have an important role to play in facilitating reduced emissions, yet the contribution of urban areas to emissions is often unclear (UN-Habitat, 2011, p. 33). There are several explanations responsible for prevailing gaps in knowledge, and these can be classified into four broad categories: conceptual gap, empirical gap, methodological gap and policy-governance gap. Conceptual gap arises from the group of issues that inhibit a clear theoretical understanding of how nations, settlements or societies, when they develop and urbanize, influence their carbon footprint. Empirical gap comes from either lack of sufficient scientific causations about urban carbon footprints or evidential data to support the prevailing theoretical knowledge. Methodological gap is caused by a gamut of reasons that inhibit proper assessment of a city's GHG emissions. Policy-governance gap is a consequence of limited knowledge with cities about their roles and responsibilities to effectively respond and act in climate mitigation. Different knowledge gaps prevailing in the interdisciplines of cities and climate change are summarized in Table 1.2, followed by a discussion.

### *1.7.1 Conceptual gap: understanding of carbon footprint*

There seems to be a preoccupation of research on discourse on the interrelationship between population, economy/affluence to emissions generally during internation comparability, at the national scale, and very recently for some global cities.

(a) One of the fallacies of such a conceptualization is that the population very often reduces most of the knowledge on carbon footprints of countries and cities to aggregated calculations. Within cities, population is studied in terms of size and its distribution (i.e. density). Urban density and spatial organization are crucial elements that influence energy consumption, especially in transportation and building systems (World Bank, 2010). Throughout, there is a recognition that changing densities in urban centres can both affect, and be affected by, global environmental change (Dodman, 2009b). But overenthusiasm of reducing

*Table 1.2* Summary of prevailing knowledge gaps

| Conceptual gap | Empirical gap | Methodological gap | Policy-governance gap |
|---|---|---|---|
| Preoccupancy with IPAT-based theory linking climate impacts and causes with population, affluence (GDP/income) | Lack of reliable data or measurement of city emissions | Differences in a consistent definition of urban | Gap in understanding in how to influence or affect low-carbon urban policy and governance |
| Lack of longitudinal research on thresholds of development/ urbanization | Lack of sufficient data on urban spatial parameters, master plans, etc. | Variable methods to allocate responsibilities – production vs. consumption criteria | Lack of integration with climate adaptation and resilience in cities |
| Lack of cross-sectional or integrated perspective considering land use, urban spatial parameters | | Difference in available inventories, tools and methodologies and their incomparability | |

national or city emissions to average values (represented as per capita $CO_2e$) for the purposes of estimation and comparison overlooks the cumulative or mass effect of these GHGs. It falls short in seeing the larger picture, where several settlements within a nation or region could collectively create a snowballing effect of harmful emissions.

There could be several implications to this kind of an approach. Highly populated cities could exhibit smaller carbon footprints on an average (per capita) basis, and vice versa is also possible, some examples being New York, London and Tokyo. The converse is also possible – two settlements with similar footprints could have completely different population sizes, urbanization or urban growth patterns, and all those factors that create a city. For instance, in Table 1.3, Denver (21.5 t/capita) and Sydney (20.3 t/capita) have similar average emissions, but they are very different in their geophysical conditions, size or land area, energy demand, economic activities, land use and transportation networks. There are similar spatial variations between Shanghai (11.7 t/capita) and Toronto (11.7 t/capita), and Barcelona (4.2 t/capita) and Seoul (4.1 t/capita).

**(b)** Similarly, there is a rich and ever-growing body of literature in parallel that informs us about the strong association between economic indicators, such as income and affluence, with carbon emissions. Heil and Wodon (1997) use the Gini index to measure the inequality of per capita emissions across different countries and the relative contribution of two income groups (poor and rich countries) to this inequality. They later employ this methodology for analysing future inequality in per capita emissions using business-as-usual projections to

*Table 1.3* Top 15 carbon-emitting cities arranged in decreasing order (of per capita emissions), showing contradictions and limitations of using population as a conceptual and empirical tool

| City | Rank | Per capita emissions (in tons) | National average (in tons) | City population (approximate figures) |
|---|---|---|---|---|
| Rotterdam | 1 | 29.8 | 12.7 | 600,000 |
| Denver | 2 | 21.5 | 23.6 | 600,000 |
| Sydney | 3 | 20.3 | 25.7 | 4,500,000 |
| Stuttgart | 4 | 16.0 | 11.6 | 600,000 |
| Shanghai | 5 | 11.7 | 3.4 | 19,200,000 |
| Toronto | 6 | 11.6 | 22.6 | 5,500,000 |
| Bangkok | 7 | 10.7 | 3.8 | 11,900,000 |
| New York | 8 | 10.5 | 23.6 | 19,000,000 |
| London | 9 | 9.6 | 10.5 | 12,000,000 |
| Cape Town | 10 | 7.6 | 9.9 | 35,000,000 |
| Tokyo | 11 | 4.9 | 10.8 | 35,200,000 |
| Barcelona | 12 | 4.2 | 9.9 | 1,600,000 |
| Seoul | 13 | 4.1 | 11.5 | 25,000,000 |
| Stockholm | 14 | 3.6 | 7.1 | 2,000,000 |
| São Paulo | 15 | 1.4 | 4.2 | 20,100,000 |

Source: Adapted from IIED (2011)

the year 2100 (Heil and Wodon, 2000). One of the main findings is that both emissions and income inequalities have decreased during past decades, which is later supported by Padilla and Serrano (2006). They further emphasize that "If you belong to the same income group (per capita), emission inequality will be less profound", especially in the low-middle-income group. Duro and Padilla (2006) applied the decomposable Theil index by decomposing Kaya identity and found that income per capita, or simply put affluence, is the main driver of emission inequality, although differences in energy intensity, and in carbon intensity of energy, were also relevant.

---

**Box 1.5   Criticisms of allocating responsibility of GHGs on average national per capita emission figures**

It is known that allocating responsibility for GHG emissions through average per capita emissions figures for nations is misleading for at least two reasons. The first is that these figures are based on where GHGs are emitted, and not on what caused them to be emitted. If GHG emissions were allocated to the location of the consumers whose consumption was the root cause of these emissions, it would considerably increase the GHG emissions per person in most high-income nations (and cities) and considerably decrease the GHG emissions per person in nations (and cities) that were successful exporters of consumer goods (especially those with high GHG emissions caused by their manufacture and transport to markets). The second is that it is very misleading to discuss responsibility for GHG emissions per person using national averages because of the very large differences in per capita emissions within each nation between the highest-income and lowest-income groups – perhaps a 100-fold or more difference between GHG emissions per person if the wealthiest 1% and the poorest 1% in many nations could be compared (Satterthwaite, 2009, p. 59).

There is also a complex series of interactions among urban density, economic status and greenhouse gas emissions. Residents of the densely populated cities of low- and middle-income countries are generally wealthier than residents of the hinterlands, yet they are far less wealthy than residents of the less densely populated cities in high-income countries. This illustrates that the relationship between urban density and greenhouse gas emissions is not straightforward. In low-income countries, residents of denser settlements are likely to have higher per capita emissions as a function of their greater wealth than residents of surrounding areas, whereas in high-income countries, residents of denser settlements are likely to have lower per capita emissions than residents of surrounding areas as a result of smaller housing units and greater use of public transportation systems (Dodman, 2009b, p. 76).

In addition, differences in per capita incomes between the richest and poorest persons, in both rich and poor countries, are bigger than the differences in per capita incomes between countries. But if all countries in fact had completely egalitarian income distributions, there would still be huge income differences based on the gaps in national incomes (Milanovic, 2005).

The research in this area underscores that global policies seeking equitable access and allocation of atmospheric space would be required to deeply reduce rooted income inequalities between the national economies (developed and developing countries) and within them, in addition to expanding technical improvements in the energy intensity and carbon intensity of energy. This is reinforced by another set of findings that challenge our conventional principles to approach issues within the development and environmental complex. It establishes that, unlike other environmental pollution, the environmental Kuznets curve (a hypothesized relationship between environmental quality and economic development: various indicators of environmental degradation tend to get worse as modern economic growth occurs until average income reaches a certain point over the course of development, thereafter improving; represented by an inverted u-shaped curve) is not the most appropriate way to explain global warming against business-as-usual development, and appropriate policies are required to reach a turning point in the relationship between income and emissions (Cantore and Padilla, 2009).

Furthermore, Groot (2010) exhibits that standard tools in the measurement of income inequality, such as the Lorenz curve and the Gini index, can successfully be applied to the issues of inequality measurement of carbon emissions and the equity of abatement policies across countries. Meanwhile, this preoccupation with population and economic parameters to understand the carbon footprint of nations and cities stems from the IPAT model, which is considered the most fundamental model applied to study environmental impacts. IPAT basically means that human impact ($I$) on the environment equals the product of population ($P$), affluence ($A$) and technology ($T$). It describes how our growing population, affluence and technology contribute towards our environmental impact. As the above discussion shows, it falls short of explaining for many emissions that have physical or spatial associations (see Figure 1.7).

But there exists no clear knowledge of the process. When nations in their different stages of development transform from rural to urban, how do their carbon footprints change? Many researchers and policymakers in fact possess an indifferent attitude towards not just this debate, but urbanization in general. For instance, many countries, including India, do not have a national urban/urbanization policy. They rather assume that since cities, unlike industrial or thermal power plants, do not directly emanate GHGs, they hardly contribute to global carbon emissions. The argument gets complicated at times, because cities themselves are a diverse lot and it is difficult to paint all of them with the same brush. In addition, many cities with tertiary economies actually have very low per capita emissions. But it is important not to miss their overall throughput in disguise of the low per capita emissions they demonstrate because of agglomeration of people or scale of economies.

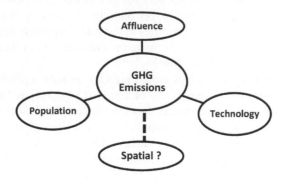

*Figure 1.7* The IPAT model and its missing link with spatial parameters

**(c) Lack of longitudinal research: The pace of urbanization and its thresholds:** Human settlements are not static in time. Urban societies and their systems, infrastructure, etc. evolve, stabilize and at times degenerate. A sound understanding of urban development pathways and their GHG consequences is needed for various urban typologies. Such an understanding can be generated from a set of well-coordinated and comparative studies across cities. Global scientific programmes can help to foster research protocols and facilitate such comparative studies (Dhakal, 2010). If we are to genuinely assess the impact of cities upon a macro-phenomenon such as climate change, a comprehensive assessment for the pace of urban development or urbanization, settlement strategy, etc. at the macro level is paramount.

There is a growing consensus among the experts and within the present scientific knowledge for the need to design or redesign cities that are environmentally friendly and resilient to climate vulnerabilities (Kropp and Reckien, 2009). Furthermore, there is an urgent need to set targets (i.e thresholds for development). There are emission targets for some countries, but none of the existing literature gives a comprehensive model towards physically shaping the futuristic built environment, based upon the potential threshold of cities' GHG emissions. This kind of forecast essentially requires a thorough understanding of our cities, their growth trajectories and how they influence climate change. Hence, it becomes important to first acknowledge the urbanization status of a country. Urban units within a country collectively represent not just a booming population and economy, but are fundamental units of local governance too. Furthering research in this area is important to guide national and international policy and governance on urbanization, environment and global change.

**(d) Lack of cross-sectional research: Integrate perspective:** As discussed earlier, the present global, national and citywide assessments are based on population, which is not a direct or complete determinant of emissions in human settlements. There is also a vast discourse to establish links between carbon emissions and the economy of settlements (World Bank, 2010), yet the spatial perspective of human settlements in emission assessment is not evident, be it at the global, national or local level. For example, take the case of land use planning in cities. It is observed that land use regulation/development controls in cities triggers growth elsewhere – in suburbs or peri-urban areas – further raising carbon emissions, highlighting the importance of land use and zoning regulations (Glaeser and Kahn, 2010). Thus, rational and scientific land use planning of the city and its region emphasizes the importance of integrated development.

The literature suggests that although there is broad agreement that the energy-efficiency improvement in urban transport and building sectors has a large potential to address carbon mitigation (Dhakal, 2010), a key gap remains in analyses involving the opportunity to optimize the urban system as a whole in an integrated fashion. Going beyond the conventional carbon management opportunities in cities, few studies have explored, separately, the opportunities to sequester carbon in urban systems too. An integrated analysis not only addressing the efficiency gains in traditional sectors, but also covering all infrastructure, technology, urban design and sinks (noting their other ecosystem services as co-benefits too), can provide a comprehensive picture of carbon management opportunities.

### 1.7.2 Empirical gap: spatial causations and their evidence

Improper measurement and management of GHGs acts as a huge gap existing between science and policy. It is believed that, irrespective of the form or timing of climate change policies, the foundation of policy design and implementation measures to address the root cause of climate change is reliable metrics on GHG emissions (and removals) (Gillenwater, 2011). The only feasible way to stabilize GHG concentrations is managing the (anthropogenic) emissions, no matter what this stabilized level of concentrations would be. And, as the old saying goes, *you cannot manage what you do not measure.* Adequate measurement of GHGs and their emissions is therefore a prerequisite to managing them (Pulles, 2011).

Literature in the interdisciplines of cities and climate change duly establishes certain theoretical causations for urban-induced GHG emissions, and recognizes significant sectors and drivers such as industry, transport, urban form, municipal waste, etc., but there is a lack of complete understanding in spatial causation of these. For instance, it has been suggested that there are a variety of factors influencing the total and per capita emissions of a city, including its geographic situation (which influences the amount of energy required for heating and lighting), demographic situation (related to both total population and household size), urban form and density (sprawling cities tend to have higher per capita emissions than compact ones), and urban economy (the types of activities that take place, and whether these emit large quantities of GHGs) (UN-Habitat, 2011, p. 61), but there is a paucity of sufficient empirical data to substantiate how some of the above-mentioned urban space-related parameters affect carbon footprints of diverse urban areas, and to what extent. As such, there is a gap in understanding of a city's spatial causations to GHG emissions. GHG inventories only measure numerical emissions across various economic activities or 'sectors', and do not report on their causations, drivers or mutual associations and co-relations. This may be appropriate for country-level footprint analysis, but for a system of an urban scale, which has a larger bearing and implications due to cause–effect relationships, it is imperative to understand how various urban spatial parameters influence a city's footprint and what could be the mutual associations between various sub-sectors.

It has been argued that a better understanding and quantification of GHG emissions and mitigation potential under the different definitions of urban areas and allocation principles globally and nationally is necessary (Dhakal, 2010). The effectiveness of any effort to address climate change relies on the availability of transparent, consistent, comparable, complete and accurate (TCCCA) data on past trends, as well as projections of GHG emissions. This goes beyond the arithmetic of accounting for carbon atoms, and clearly relates to legal, economic and behavioural aspects, making the field of measurement and management of GHGs an interdisciplinary field, linking natural, economic and behavioural sciences with procedures of law and regulation. Thus, addressing this empirical research gap is bound to influence better mitigation policies, climate co-benefits associated with urban planning, land use planning, industrialization, transportation, waste management, etc., and their enhanced integration with urban governance.

### 1.7.3 Methodological gap: estimating carbon footprint

There has been an increasing debate about the proportion of global emissions that can or should be attributed to urban areas (or a city's carbon footprint). This is inherently because of three main anomalies: (a) urban definitions; (b) allocating responsibilities; and (c) available methodologies and tools.

**(a) Urban definitions:** Methodological differences in urban footprinting are compounded by variations in the definition of 'urban areas' used by different countries and the ways in which urban boundaries are defined (for various definitions of urban and the quality of data available, see UN-Habitat, 2011, p. 61). The United Nations defines an urban agglomeration as the built-up or densely populated area containing the city proper, suburbs and continuously settled commuter areas. It may be smaller or larger than a metropolitan area; it may also comprise the city proper and its suburban fringe or thickly settled adjoining territory. A metropolitan area is the set of formal local government areas that normally comprise the urban area as a whole and its primary commuter areas. A city proper is the single political jurisdiction that contains the historical city centre. However, an analysis of countries worldwide shows that different criteria and methods are being used by governments to define urban:

- 105 countries base their data on administrative criteria, limiting it to the boundaries of state or provincial capitals, municipalities or other local jurisdictions; 83 use this as their sole method of distinguishing urban from rural.
- 100 countries define cities by population size or population density, with minimum concentrations ranging broadly from 200 to 50,000 inhabitants; 57 use this as their sole urban criterion.
- 25 countries specify economic characteristics as significant, though not exclusive, in defining cities – typically, the proportion of the labour force employed in non-agricultural activities.

- 18 countries count the availability of urban infrastructure in their definitions, including the presence of paved streets, water supply and sewerage systems or electricity.

As this research is based primarily in the context of India, the definition of urban widely followed in India (i.e. the Census of India) has been adopted. The Census of India (2011) recognizes all those settlements as urban which either have a statutory status such as municipal committee/corporation/notified area committee/cantonment board, estate office, etc., or that are *census towns* that fulfil all of the following three conditions simultaneously: (1) a population of more than 5,000; (2) more than 75% of the male working population engaged in non-agricultural activities; and (3) a density of population of more than 400 people/sq km. Accordingly, India had an urban population of 377 million in 2011, which is 31.16% of the total population (as against 27.81% in 2001), distributed across 7,935 towns, of which 4,031 are statutory towns and 3,894 are *census towns*.

   (b) **Allocating responsibilities:** Urban areas also behave as a complex and dynamic entity between nation and individuals with unclear functional/physical boundaries, consumption and production issues, scope of measurement – direct, indirect emissions, etc., which has led to an absence of consensus on a single methodology, inventory or tool to be followed worldwide. An assessment of the contribution of urban centres or urbanization (growth in the proportion of a national population living in urban areas) to climate change can be done either from the perspective of "where GHGs are produced" (by assessing what proportion of GHGs emitted by human activities comes from within the boundaries of urban centres) or from the consumption perspective, assessing all the GHGs emitted as a result of the consumption of resources and waste generation by urban populations, no matter where they originated (Satterthwaite, 2009, p. 45). From the production perspective, if energy-intensive production is concentrated in cities, their average GHG emissions per person will increase (unless the production is served by electricity not generated from fossil fuels). This infers that cities, particularly in low- and middle-income nations with heavy industry or fossil-fuelled power stations, can have very high $CO_2$ emissions both on a cumulative and per capita basis. But in many countries, a considerable proportion of energy-intensive production (for instance, mines and mineral processing) or fossil-fuelled electricity generation takes place in rural areas or urban areas that are too small to be considered cities in the conventional sense. Rural districts with such energy-intensive production can have per capita GHG emissions that are much higher than most cities – although most city GHG emissions inventories that use the production perspective still employ a consumption perspective with regard to electricity (wherein the emissions generated by the electricity used in the city are allocated to the city, not to the location where the electricity was generated).

   In addition, when comparisons for GHG emissions are made between rural and urban areas, where the high contribution of urban areas is stressed, generally, no consideration is given to emissions from agriculture and land use changes in rural areas. The IPCC suggests that the latter accounts for a major portion of all human-induced GHG emissions (IPCC, 2014). One obvious objection to using the production approach is that a large proportion of the products of rural-based mines, forests and agriculture, as well as land use changes, are meant to serve the production or consumption needs in urban areas. Therefore, it is misleading to allocate these to rural areas or rural populations. But the real issue here is the inappropriateness of allocating responsibility for GHG emissions to a nation as a whole (and by implication to that nation's entire population) or to urban areas in general or to particular cities (and by implication to all the urban population or to the populations of particular cities). Human-induced GHG emissions are not caused by 'people' in general, but by specific human activities by specific people or groups of people.

(c) **Available methodologies and tools:** Similar to downscaling actual data, climate studies employ spatially disaggregating methods on demographic projections and scenarios in recent global assessments. For instance, various models for downscaling the IPCC's Special Report on Emissions Scenarios (SRES) scenarios to the grid level have been developed (Gaffin *et al.*, 2004; Grübler *et al.*, 2007; van Vuuren *et al.*, 2007). These models downscaled drivers of climate change: emissions, population and GDP. In the most recent of the examples cited (van Vuuren *et al.*, 2007), 17 world regions of the SRES scenarios were downscaled to the national- and grid-level ($0.5° \times 0.5°$) resolution (i.e. 55 km × 55 km at the equator). In order to achieve the grid estimation of population, national growth rates for each of the SRES scenarios have been linearly applied to each grid cell of the Gridded Population of the World (GPW 2000) data, which are used as the base grid (SEDAC/CIESIN, n.d.). The gridded GDPs were obtained by multiplying the national GDP per capita by the population grids. GHG emissions were only disaggregated at the national level by means of the $I = PAT$ model. For further subregional or national studies, finer downscaling might be needed in order to reach decision-making relevance at corresponding scales. In particular, emissions should be disaggregated at the grid level, and more data on land cover/land use change incorporated (Dao and van Woerden, 2009, p. 231).

On the other hand are aggregation-based methods, where urban development scenarios at present, linked with an agent-based urban metabolism model, have been used to demonstrate the potential to build (and rebuild) existing cities in ways that will extract much less specific energy and material resources from the environmental system, thereby reducing GHGs and improving the climate change performance of urban systems (Schremmer and Stead, 2009). Fundamentally, all the prevailing methodologies and tools are derived from the Kaya identity. It is an equation conceptually based on the IPAT model relating factors that determine the level of human impact on climate, in the form of emissions of carbon dioxide:

Kaya identity: $CO_2 = (CO_2 / TOE) \times (TOE / GDP) \times (GDP / POP) \times POP$

i.e. $CO_2$ emissions = carbon content of energy × energy intensity of the economy × production per person × population

**Essentially, GHG emissions = activity data (fuel consumed) × emission factor**

While 'activity data' is influenced by various spatial factors within the city such as its scale/size, geographical and climatic location, spatial structure or form, land use and transport interactions, assimilative capacity of open/green areas, to name a few, there is no methodology or model that measures urban spatial parameters and assists urban planners to assess a city's carbon footprint. Several models, while downscaling IPCC's global SRES scenarios on population, GDP and GHGs to a subnational or city context, offer limited accuracy and precision. Moreover, most of the methodologies and tools in practice have been formed through study and evidence of cities in developed countries, particularly in Europe and the US. Hence, there is a need to develop contextual and specific methodologies in the national contexts. A complete discussion on GHG methodologies/inventories and major inconsistencies therein has been dealt with in Chapter 3.

In general, a large volume of literature on embodied energy has been published using input–output analyses, material flows, and life cycle assessment at national, sectoral and product levels, but application to study the carbon footprint of cities is less (Dhakal, 2010). Whatever inventories or tools that exist have been developed by energy experts that, as a rule, estimate emissions on the basis of activity data and emission factors, thus making it difficult for urban policymakers, planners and practitioners to comprehend, use and apply these in their iterative thinking, planning, appraisal and decision-making process. It has also been argued that urban systems, essentially being open systems with extensive cross-boundary interactions for food, water, energy, mobility, material and services, require the development of new types of methodologies going beyond product-based life cycle assessments, input–output analyses and household income expenditure surveys. These could assist in a better understanding of the prospects for optimizing GHG mitigations with multiple benefits locally for issues such as air pollution, transport, and waste management. In addition, the mitigation and adaptation needs to be connected for urban development planning to optimize urban infrastructure systems and financial resources (Dhakal, 2010). In view of the above problems, there is an imperative to have an easy-to-use inventory or a tool for spatial planners, policymakers and decision-makers that is based on urban spatial parameters to correlate with a city's carbon footprint.

### 1.7.4 Policy-governance gap

The policy-governance gap is concerned with the overall climate policy landscape in the country and the role of city authorities in effective climate mitigation and co-benefit initiatives. Dhakal (2010) argues that a large body of literature has analysed urban carbon governance from multiple levels (Alber and Kern, 2008; Allman *et al.*, 2004; Betsill and Bulkeley, 2007; Bulkeley and Betsill, 2003; Bulkeley and Kern, 2006; DeAngelo and Harvey, 1998; Holgate, 2007; Kern and Bulkeley, 2009; Kousky and Schneider, 2003; Romero-Lankao, 2007). The literature points out that the local government can govern climate change mitigation in four ways: self-governing (reducing GHG from municipal actions and activities), governing through legislating, governing by provisioning and governing by enabling (Alber and Kern, 2008). The rising interest of local governments to assume more responsibility to govern cities seems a positive trend for urban carbon mitigation. Nevertheless, the academic literature and policy debates overemphasize the role of local government and fail to take into account the limited ability of municipal governments in reducing substantial amounts of emissions from urban activities because of several structural factors in cities (the city playing the role of facilitator rather than actor, provisioning of municipal utility services to private sectors, limited authorities, crumbling financial performance, etc.). In developing countries, the capacity, resources and jurisdiction of city governments are further limited. Thus, the role of municipal government is absolutely necessary, but not sufficient, for urban carbon management. The above suggests for contributions of other agents such as national or state governments within the multilevel framework or non-governmental organizations, civil society or scholars.

In an urban setting, municipal governments often have a limited role in managing many key drivers of even direct GHG emissions. Mobility, food and material flow are difficult to influence. For instance, in the case of India, most of the urban bodies are weak and encounter immense operation and management pressure, staff shortages, a lack of sufficient capacities, equipment and technologies, and overlapping jurisdiction to deal with elementary urban issues (Jha and Vaidya, 2011; Sethi, 2011; Singh, 2011; Siwach, 2011). Expenditure on salaries and wages accounts for 54.2% of the total municipal expenditure. In several states, however, it is as high as 80.4% (India Infrastructure Report, 2006). In addition, there are external challenges such as the growing population, horizontal coordination with other agencies, political allegiance and transparency, thus putting the mitigation of climate change on the back burner. In this context, research advances are needed in determining the efficient and innovative ways of governing urban GHGs for reducing the overall footprint.

In this context, it is important to reflect upon how local action is influenced by national strategies. Climate change finds a narrow reference in the national environmental policy framework (i.e. National Environment Policy, NEP). Most of the document (Ministry of Environment and Forests, 2006) clarifies India's position in the international climate change debate rather than offering a nationwide integrated approach on the subject. It upholds the principle of common but differentiated responsibilities and respective capabilities of different countries. Pertaining to mitigation, the policy emphasizes multilateral approaches, rights to equal per capita entitlements of global environmental resources, priority to the right to develop, and encouragement of Indian industry to participate in clean development mechanisms through capacity building.

This was followed by a national-level policy, the National Action Plan on Climate Change (NAPCC), adopted by the central government in 2008, which identifies eight missions: National Solar Mission, National Mission for Enhanced Energy Efficiency, National Mission on Sustainable Habitat, National Water Mission, National Mission for Sustaining the Himalayan Ecosystem, National Mission for a Green India, National Mission for Sustainable Agriculture and

National Mission on Strategic Knowledge for Climate Change (Ministry of Environment and Forests, 2008). Themes bearing strong potential to influence urban India are: National Missions on Sustainable Habitat, Energy Efficiency, Solar Mission, Green India and, to an extent, Strategic Knowledge for Climate Change.

There are two major concerns where the policy confines its outlook. First, it does not follow an integrated view, but a squarely sectoral approach to contain GHG emissions. Second, it is limited to the identification of the institutional and procedural mechanisms that will enable the action plan to function, so as such it is virtually a vision paper. Since there is so much variation in terms of types of GHGs, sector contributions and drivers, fuels used, their availability, pricing, market economics, social behaviour, governance structures, financing mechanisms, etc., there are bound to be a myriad of policy and governance options to be optimized for different municipal governments. Although NEP 2006 and NAPCC 2008 are the most pertaining national policies with respect to climate change, both seem to lack an integrated or cross-sectoral perspective and requisite degree of detail to deliver targeted and measurable outputs.

## 1.8 Research questions and hypothesis

Based on this extensive study of significant documents in the subject area, there are several grey zones and underpinnings on the role of cities and urbanization in contributing towards climate change. The existing knowledge may be characterized by:

- Full of discourse co-relating carbon emissions with the population and economy (GDP) of developed and developing countries.
- Lack a holistic understanding of the role of cities, which act as the economic engines of these countries in causing GHG emissions.
- Emissions are widely understood as a function of "per capita $CO_2e$" in average figures.
- It has already been established that certain spatial determinants of the city such as form, density and public transport are responsible for influencing its carbon emissions.
- There is no clear consensus on what should be the urbanization policy for the future in the wake of climate change. There is a need to extend knowledge on what should be the appropriate urbanization strategy in various geographical regions, considering their individual capacities to develop.

This analysis supports some of the ongoing research needs: the need for a better understanding of the nature of emissions from cities; the considerable differences in GHG emissions between cities and the wide range of factors contributing to this; the substantial differences in responsibility for GHG emissions from different groups of people within cities; and the importance of examining the underlying drivers of emissions (UN-Habitat, 2011, p. 61). The preliminary literature review of the present body of knowledge leads to many interrogations, such as: What should be the appropriate urban scale and density for cities? What should be the most suitable spatial pattern for a city? What is the most favourable rate of urbanization? What are the appropriate regulatory and governance mechanisms to necessitate low-carbon development? These research questions could be broadly classified according to the respective knowledge gaps they belong to, as discussed below:

**Conceptual:** Role of cities in contributing to climate change?

**Methodological:** Understanding why certain cities cause higher GHG emissions than others?

**Empirical:** Can we have an alternative mathematical equation/functional relation to ascertain GHG emissions of a city, based upon its urban spatial parameters?

**Policy-governance:** What could be future low-carbon strategies for urbanization?

It needs to be clarified that in spite of the shortcomings in approaches associated with environmental models dependent upon population, affluence (particularly GDP) and technological parameters (as in the IPAT equation), or emission accounts dependent upon energy intensity, emission factors and activity data (as shown in the Kaya identity), this research does not overrule the significance of such parameters or the models based upon them, but for more practical research on cities, it intends to explore the role of spatial parameters that could possibly better associate with the degree of activity within a city (all things being equal), their energy demand, and hence carbon footprint. Hence, this research is based on the premise or hypothesis that *urban settlements in developing countries significantly contribute to national carbon footprints, which could be better understood and measured on the basis of their 'spatial' parameters.*

In applied terms, if one aims to realize an urbanization policy and urban governance that is inherently low-carbon, it is imperative to ascertain the following:

(a) The relationship between economic development, urbanization and GHG emissions, at the global and national levels.
(b) The overall quantum of GHGs contributed by urban areas, with main activities/sectors responsible at the national level.
(c) The urban spatial causations of GHG emissions at the local city level.

This necessitates having a research plan with specific objectives and an equally responsive methodology.

## 1.9 Research plan

### *1.9.1 Research objectives*

Taking into consideration the identified gaps in knowledge (section 1.7) and interrogations relevant to the subject area (section 1.8), this research intends to study urban-induced GHG emissions in a spatial perspective – as a function of interwoven urban spatial parameters. Accordingly, the step-by-step objectives to attain this could be enumerated as the following (for visual aid, see Figure 1.8):

1 To assess the interrelationship between development, urbanization and global climate change, with special reference to GHG emissions in India.
2 To evaluate the cumulative GHG emissions/carbon footprint for entire urban areas of India.
3 To study the causation of GHG emissions from urban spatial parameters (such as geographical location, size, urban form, city structure, predominant landuse, agglomeration, etc.).
4 To formulate a statistical correlation or a tool/model with a suitable measure/metric to report urban GHG emissions based on urban spatial parameters.
5 To recommend strategies that form the basis for low-carbon urbanization policy, spatial planning and governance.

### 1.9.2 Data, methodology and reporting

As explained about the methodology-related knowledge gap in section 1.7, at present there is no single methodology to empirically determine the role of

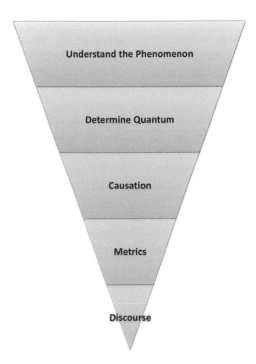

*Figure 1.8* Research objectives

urban areas in contributing to national GHG emissions, considering the variety in spatial scales one may intend to study in this relationship. Accordingly, this research undertakes a comprehensive review of existing city-based methodologies, tools and models, in addition to the ones at national/country and corporate/business/household levels to infer their influence on developing relevant urban methodologies. Upon a critical assessment of existing methodologies, the research adopts an innovative three-tier approach to empirically establish a functional relation between spatial parameters of urbanization and GHG emissions, at international, national and local levels, viz:

1   *International level*: Emissions data of 209 countries/territories (which includes data of 150 developing countries) is disaggregated using standard sort and filter operations, for all three regions – developed, developing and least developed – on the urban–rural gradient, for three categories – urban, urbanizing and rural – numerically defined as $0 < x < 33$, $34 < x < 67$ and $x > 67$, where $x$ is urban population (in percent) for a country/territory for a particular data year for all six decadal points (1960–2010), and tabulated to generate: (1) a 3D surface that shows trends in values across two dimensions (development state and urbanization, in this case) and carbon emissions as the third dimension forming a continuous three-dimensional surface; and (2) a contour that resembles a surface chart viewed from above. In both cases, colours represent ranges of values (carbon dioxide emissions in tons per capita).

2   *National level*: Meso-methods such as the disaggregation of national GHG data to quantify urban contribution of GHGs. It is a form of spatial numerical analysis, whereby similar techniques have been employed in assessing global urban GHGs or emission contribution of European and Australian cities.

3   *Local/city level*: Mathematical modelling of seven urban spatial parameters and corresponding indicators of 41 Indian cities to establish a correlation with their GHG data using OLS regression analysis in Stata 2012 software. In order to pilot-test the coefficient of correlation, its positive/negative sign and nature of equation (linear, polynomial, logarithmic and exponential), an inventory of selected Indian cities is prepared and correlation analysis conducted over 10 basic indicators of the above seven parameters. The city selection was based on stratified samples across diverse criteria, which included regional climate, functional diversity (in terms of economic activity), urban hierarchy, variation in urban form and spatial structure.

The materials and methods employed in the research, for each of its objectives, are summarized in Table 1.4. The respective reporting of the analysis, results and discussion is also stated.

*Table 1.4* Objectives, data and methods and reporting of the research

| Objectives | Data and methods | Reporting |
|---|---|---|
| To assess the interrelationship between development, urbanization and global climate change, with special reference to GHG emissions in India. | • Literature study<br>• Time-series analysis using secondary data<br>• Urbanization data: Census of India (2011)<br>• Emissions data: National Communications to UNFCCC by Government of India (MoEF, 2012); economic data from Central Statistical Organization | Chapter 2 |
| To evaluate the cumulative GHG emissions/carbon footprint for entire urban areas of India. | • Review of important methodologies, tools and inventories on carbon footprinting/ calculating GHG emissions<br>• Emission data from national communications to UNFCCC by Government of India (MoEF, 2012) | Chapter 3 |
| To study the causation of GHG emissions from urban spatial parameters (such as geographical location, size, urban form/density, city structure, predominant land use, agglomeration, etc.) | • Theoretical causation from literature study<br>• To select two or three sample cities of diverse size and specialization in different geographical locations<br>• Spatial data: spatial attributes from respective city development/master plans, secondary data sets from line departments and primary surveys to collect data on LU and networks.<br>• GHG data: ICLEI's inventory for emissions<br>• Pilot study applying correlation analysis for each parameter | Chapter 4 |
| Formulate a statistical correlation or a tool/ model with a suitable measure/metric to report urban GHG emissions based on urban spatial parameters. | • To formulate a functional relation/ mathematical equation using OLS regression modelling to internalize diverse spatial parameters that have a strong correlation with GHG emissions<br>• To test the model, validate results of sample cities with overall city emission data and appropriate the model if necessary | Chapter 5 |
| Recommend strategies that form the basis for low-carbon urbanization policy, spatial planning and governance. | • Discuss the effectiveness of studying and reporting urban GHG emissions from the perspective of its spatial parameters<br>• Deduce inferences from Chapters 2, 3, 4 and 5 to suggest strategies for low-carbon future pathways | Chapter 6 |

### 1.9.3 Data analysis and techniques to be applied

The research is based on intensive application of analytical and quantitative techniques at multi-spatial scales, such as:

*Global*: 3×3 spatial development matrix, time-series analysis

*National urban–rural*: spatial disaggregation methods

*Local urban*: statistical co-relation and OLS regression modelling

*Softwares*: Stata 2012, AutoCAD 2010, MS Excel 2010, MS Word 2010, MS PowerPoint 2010

*Pilot Study*: Subsequent to the preparation of a model/matrix, a pilot study shall be undertaken in selected Indian cities of different features to evaluate/ test the validity of the hypothesis. The sampling shall be based on keeping in view variations of geographical location, population and size of city, socio-economic conditions, predominant economic activity, function/land use, city form/structure, etc. For details, see Chapter 5.

## 1.10 Salient features and limitations of the study

This study is the first of its kind in multiple ways. There is no single study across the globe that has tried to understand a city's carbon/GHG footprint from the perspective of a physical/spatial planner, as a consequence of growing urbanization or spatial parameters, based on their causation, co-relation and supported by empirical data. The challenge of this research could be gauged from the fact that this area of research is in its incipient stage in most developing countries such as India that have limited data sets on many urban and emission indicators in comparison to developed countries, particularly at the city level. The research attempts to address this by conducting analysis at demonstrable and multiple spatial scales, namely at the national urban and city levels within the methodology.

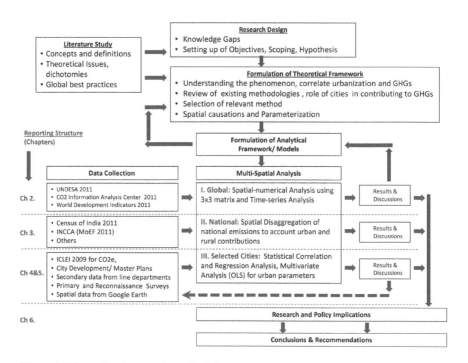

*Figure 1.9* Procedural research methodology

The sampled cities have also been carefully selected to represent the diversity of location, geophysical conditions, city size, etc. within the urban–rural hierarchy. Based on the research objectives and the scope of the study, the research methodology is mandated to be applied, descriptive, analytical, quantitative and empirical. Salient features are briefly outlined below:

(a) *Applied research*: Attempts to find a solution to practical and immediate problems being faced by society.
(b) *Descriptive research*: Attempts to explain a phenomenon – the impact of cities on climate change.
(c) *Analytical research*: Utilizes facts and information already available to analyse and make critical evaluations.
(d) *Quantitative research*: Based on parameterization and measurements of variables.
(e) *Empirical research*: The research is evidence-based and can be further corroborated.

The research contributes to the theory in the form of research and policy papers, in national and international publications. It inputs practice by providing new methods/tools for spatial planners to rationally and equitably account for carbon footprint. In addition, it recommends an urban governance framework inclusive of a green/low-carbon agenda.

In spite of the above merits, there are certain limitations to this kind of study. Since the methodology is based on multivariate analysis of several parameters, it is practically impossible to physically collect primary data on citywide indicators for several cities in the shortest possible time. Hence, certain data pertaining to city indicators and their carbon emissions are collected from secondary sources of information. In order to reduce anomalies, the data of each category are procured from the same source/agency and are thoroughly confirmed for consistency and accuracy.

---

## Basic questions

1   What are standard (UNFCCC and IPCC) definitions of climate change? How do they mutually differ?
2   What constitutes the family of GHGs and what is the greenhouse effect?
3   Define carbon footprint. Based on the concept of $CO_2e$ and GWP, how is it relatable to GHG footprint?
4   Differentiate between urban, urban development, urban growth and urbanization.
5   Explain how the concept of carbon footprint has evolved from the theory on ecological footprint.
6   Discuss how global (and national) climate change policy and research have evolved in the last two decades.
7   What is the cause of climate change? Explain major gaseous and sectoral contributors.
8   Under what circumstances is the carbon footprint equal to the GHG footprint of a country, city or organization? Is it possible to equate carbon footprint with ecological footprint?

**Advanced questions**

1   How do contributions to and impacts of climate change vary spatially? Is there an evident pattern or notable feature to this phenomenon?
2   Why is it important to study the spatial pattern of climate change? Substantiate it on variations in contributions to and impacts of climate change.
3   Based on normative understanding, what is the role of urban areas in contributing to climate change?
4   Enumerate major knowledge gaps in understanding the role of urban areas contributing to climate change. Discuss in detail any two significant aspects.
5   What are fallacies in using the IPAT model and the per capita emissions approach while deciphering carbon footprints from cities?
6   Argue in detail the methodological gaps in understanding the role of urban areas contributing to climate change.
7   What are the criticisms of allocating responsibility of GHGs on average national per capita emission figures?
8   How do your national policies favour or overlook a 'scientific approach' in appreciating a city's GHG emissions?
9   What is the imperative to have a multilevel research framework to account for GHG emissions from urban areas of a country?

**Do it yourself exercises**

1   Using Table 1.1 as a reference, prepare a timeline of climate change research in your country showing and highlighting key science and policy milestones.
2   Download a calculator from the Internet that helps estimate your personal/ household carbon footprint. Calculate your weekly, monthly and annual carbon footprint (there are certain activities that you do not do weekly but annually, such as long air/train travel, etc.). Compare your annual footprint with that of your city, nation and the world.
3   Make a weekly plan to reduce your individual carbon footprint. Act on it and take daily recordings for seven days. Reassess your new footprint against your earlier weekly footprint (calculated in the above exercise). Try to note down probable gaps that affect accurate estimation.
4   Start collecting baseline data on GHG emissions, gases and sector contributions, population, urbanization, energy sources, etc. for your country from standard sources/agencies. Calculate the national average GHG/carbon dioxide emissions per capita yourself and compare it with your footprint in exercises 2 and 3 above.
5   Try to find out your national environment, climate, energy and urbanization plans and policies. Evaluate how much they interact with each other's challenges.
6   Select any four global cities. Find their population, total GHG emissions and per capita GHG emissions. Compare the results.

# Suggested reading

## *On conceptual understanding between cities and climate change*

1 Satterthwaite, D. (2009). The implications of population growth and urbanisation for climate change. In J.M. Guzman, G. Martine, G. McGranaghan, D. Schensul and C. Tacoli (Eds.), *Population dynamics and climate change* (pp. 45–63). New York/London: UNFPA & IIED.
2 UN-Habitat (2011). *Cities and climate change: Global report on human settlements.* London/Washington, DC: Earthscan & UNCHS.
3 World Bank (2010). *Cities and climate change: An urgent agenda.* Washington, DC: IRDC.

## *On aspects of urban mitigation and carbon footprints*

1 OECD (2009). *Cities, climate change and multilevel governance.* OECD Environmental Working Papers No 14. Paris: OECD Publishing, pp. 30–44.
2 Dhakal, S. (2010). GHG emissions from urbanization and opportunities for urban carbon mitigation. *Current Opinion in Environmental Sustainability, 2,* 277–83.
3 GEA (2012). *Global energy assessment: Towards a sustainable future.* Cambridge/New York: Cambridge University Press & IIASA, Austria.
4 IPCC (2014). *Climate change 2014: Synthesis report. Contribution of working groups I, II and III to the fifth assessment report of the Intergovernmental Panel on Climate* [Core Writing Team, R.K. Pachauri and L.A. Meyer (Eds.)]. Geneva: IPCC.

# References

Aalborg (1994). *The Charter of European Cities and Towns Towards Sustainability.* Retrieved from http://ec.europa.eu/environment/urban/pdf/aalborg_charter.pdf on 22 November 2015.
Alber, G. and Kern, K. (2008). Governing climate change in cities: Modes of urban climate governance in multi-level systems. In Proceedings of OECD International Conference on *Competitive cities and climate change.* Milan, Italy, 9–10 October 2008.
Allman, L., Fleming, P. and Wallace, A. (2004). The progress of English and Welsh local authorities in addressing climate change. *Local Environ, 9*(3), 271–83.
Betsill, M. and Bulkeley, H. (2007). Guest editorial: Looking back and thinking ahead – a decade of cities and climate change research. *Local Environ, 12*(5), 447–56.
Bulkeley, H. and Betsill, M. (2003). *Cities and climate change: Urban sustainability and global environmental governance.* London: Routledge.
Bulkeley, H. and Kern, K. (2006). Local government and the governing of climate change in Germany and the UK. *Urban Studies, 43*(12), 2237–59.
Cantore, N. and Padilla, E. (2009). Emissions distribution in post-Kyoto international negotiations: A policy perspective. In N. Ekekwe (Ed.), *Nanotechnology and microelectronics: Global diffusion, economics and policy* (pp. 1–15). Hershey, PA: IGI Global.
Census of India (2011). *Provisional Population Totals 2011.* Paper–II, 2. New Delhi: Census of India.
Chandy, L. and Kharas, H. (2014). *What do new price data mean for the goal of ending extreme poverty?* Washington, DC: Brookings Institute, May 2014.

Dao, H. and van Woerden, J. (2009). Population data for climate change analysis. In J.M. Guzman, G. Martine, G. McGranaghan, D. Schensul and C. Tacoli (Eds.), *Population dynamics and climate change* (pp. 218–38). New York/London: UNFPA & IIED.

Dodman, D. (2009a). Blaming cities for climate change? An analysis of urban greenhouse gas emissions inventories. *Environment and Urbanization, 21*(185). doi:10.1177/0956247809103016.

Dodman, D. (2009b). Urban form, greenhouse gas emissions and climate vulnerability. In J.M. Guzman, G. Martine, G. McGranaghan, D. Schensul and C. Tacoli (Eds.), *Population dynamics and climate change* (pp. 64–79). New York/London: UNFPA & IIED.

DeAngelo, B. and Harvey, L.D.D. (1998). The jurisdictional framework for municipal action to reduce greenhouse gas emissions: Case studies from Canada, the USA and Germany. *Local Environ, 3*(2), 111–36.

Dhakal, S. (2010). GHG emissions from urbanization and opportunities for urban carbon mitigation. *Current Opinion in Environmental Sustainability, 2*, 277–83.

Duro, J.A. and Padilla, E. (2006). International inequalities in per-capita $CO_2$ emissions: A decomposition methodology by Kaya factors. *Energy Economics, 28*, 170–87.

Earth Policy Institute (2013). *Carbon emissions*. Retrieved from www.earth-policy.org/indicators/C52 on 7 May 2016.

Enescot (2011) *Cumulative energy-related carbon dioxide emissions between 1850–2005 for different countries*. Retrieved from https://commons.wikimedia.org/wiki/File:Cumulative_energy-related_carbon_dioxide_emissions_between_1850-2005_for_different_countries.png on 5 June 2016.

Gaffin, S.R., Rosenzweig, C., Xing, X. and Yetman, G. (2004). Downscaling and geo-spatial gridding of socio-economic projections from the IPCC Special Report on Emissions Scenarios (SRES). *Global Environ. Change A, 14*, 105–23. doi:10.1016/j.gloenvcha.2004.02.004.

GEA (2012). *Global energy assessment: Towards a sustainable future*. Cambridge/New York: Cambridge University Press & IIASA, Austria.

Ghosh, A. (2016, 5 June). Nobody creates a narrative around climate change. *Hindustan Times*, New Delhi. Retrieved from www.hindustantimes.com/india-news/indian-artistes-need-to-work-on-the-climate-change-narrative-amitav-ghosh/story-kQuDwZpFWUE0EN51xoXz0H.html on 7 June 2016, citing Ghosh, A. (2016). *The Great Derangement*. Gurgaon, India: Allen Lane in Penguin Books India.

Gillenwater, M. (2011). Filling a gap in climate change education and scholarship. *Greenhouse Gas Measurement & Management, 1*, 11–16. doi:10.3763/ghgmm.2010.0012.

Glaeser, E.L. and Kahn, M.E. (2010). The greenness of cities: Carbon dioxide emissions and urban development. *Journal of Urban Economics, 67*(3), 404–18.

Groot, L. (2010). Carbon Lorenz curves. *Resource and Energy Economics, 32*, 45–64.

Grübler, A., O'Neill, B., Riahi, K., Chirkov, V., Goujon, A., Kolp, P., Prommer, I., Scherbov, S. and Slentoe, E. (2007). Regional, national, and spatially explicit scenarios of demographic and economic change based on SRES. In K. Riahi and N. Nakicenovic (Eds.), Greenhouse gases: Integrated assessment (Special Issue). *Technological Forecasting and Social Change, 74*(7).

Heil, M.T. and Wodon, Q.T. (1997). Inequality in $CO_2$ emissions between poor and rich countries. *Journal of Environment and Development, 6*, 426–52.

Heil, M.T. and Wodon, Q.T. (2000). Future inequality in $CO_2$ emissions and the impact of abatement proposals. *Environmental and Resource Economics, 17*, 163–81.

Holgate, C. (2007). Factors and actors in climate change mitigation: A tale of two South African cities. *Local Environ, 12*(5), 471–84.

IIED (2011). *Cities with world's highest and lowest carbon emission*. Retrieved from www.rediff.com/money/slide-show/slide-show-1-cities-with-worlds-highest-and-lowest-carbon-emission/20110414.htm#3 on 25 April 2012.

India Infrastructure Report (2006). Urban Infrastructure. In *India infrastructure report*. New Delhi: Indian Development & Finance Corporation.

IPCC (2007a). *Climate change: Synthesis report to the fourth assessment report of the Intergovernmental Panel on Climate.* Geneva: IPCC.

IPCC (2007b). Annex I (Glossary). *Climate change: Synthesis report to the fourth assessment report of the Intergovernmental Panel on Climate.* Geneva: IPCC.

IPCC (2007c). *Climate change 2007: The physical science basis. Contribution of working group I to the fourth assessment report of the Intergovernmental Panel on Climate Change* [S. Solomon, D. Qin, M. Manning, Z. Chen, M. Marquis, K.B. Averyt, M. Tignor and H.L. Miller (Eds.)]. Cambridge/New York: Cambridge University Press.

IPCC (2014). *Climate change 2014: Synthesis report. Contribution of working groups I, II and III to the fifth assessment report of the Intergovernmental Panel on Climate* [Core Writing Team, R.K. Pachauri and L.A. Meyer (Eds.)]. Geneva: IPCC.

Jha, G. and Vaidya, C. (2011). Role of urban local bodies in service provision and urban economic development: Issues in strengthening their institutional capabilities. *Nagarlok, 43*(3), 1–38.

Kern, K. and Bulkeley, H. (2009). Cities, Europeanization and multi-level governance: Governing climate change through transnational municipal networks. *Journal of Common Market Studies, 47*(2), 309–32.

Kousky, C. and Schneider, S.H. (2003). Global climate policy: Will cities lead the way? *Climate Policy, 3,* 359–72.

Kropp, J.P. and Reckien, D. (2009). Cities and climate change: What options do we have for a safe and sustainable future? Paper presented at the 5th Urban Research Symposium *Cities and climate change: Responding to an urgent agenda,* 28–30 June, Marseille, France.

Milanovic, B. (2005). *Worlds apart: Measuring international and global inequality.* Princeton, NJ: Princeton University Press.

MoEF (2006). *National environment policy.* New Delhi: Ministry of Environment and Forests, Government of India.

MoEF (2008). *National action plan for climate change.* New Delhi: Ministry of Environment and Forests, Government of India.

MoEF (2012). *India: Greenhouse gas emissions 2007.* New Delhi: Indian Network for Climate Change Assessment, Ministry of Environment & Forests, Government of India.

Munich Re (2011). *Topics geo natural catastrophes 2010: Analyses, assessments, positions.* Retrieved from http://bit.ly/i5zbut on 23 April 2014.

OECD (2009). *Cities, climate change and multilevel governance.* OECD Environmental Working Papers No 14. Paris: OECD Publishing, pp. 30–44.

Padilla, E. and Serrano, A. (2006). Inequality in $CO_2$ emissions across countries and its relationship with income inequality: A distributive approach. *Energy Policy, 34*(14), 1762–72.

Peters, G.P. (2010). Carbon footprints and embodied carbon at multiple scales. *Current Opinion in Environmental Sustainability, 2,* 245–50.

Rees, W.E. (1992). Ecological footprints and appropriated carrying capacity: What urban economics leaves out. *Environment and Urbanization, 4*(2), 121–30. doi:10.1177/095624789200400212.

Pulles, T. (2011). Greenhouse gas measurement & management: Why do we need this journal? (Editorial). *Greenhouse Gas Measurement & Management, 1,* 4–6.

Revi, A. (2008). Climate change risk: A mitigation and adaptation agenda for Indian cities. *Environment and Urbanization, 20*(1), 207–30.

Rogner, H.H., Zhou, D., Bradley, R., Crabbé, O., Edenhofer, O., Hare B., Kuijpers, L. and Yamaguchi, M. (2007). 'Introduction'. In B. Metz, O.R. Davidson, P.R. Bosch, R. Dave and L.A. Meyer (Eds.), *Climate change 2007: Mitigation, contribution of* working group III to the fourth assessment report of the Intergovernmental Panel on Climate. Cambridge/New York: Cambridge University Press, pp. 95–116.

Romero-Lankao, P. (2007). How do local governments in Mexico City manage global warming? *Local Environ, 12*(5): 519–35.

Satterthwaite, D. (2009). The implications of population growth and urbanisation for climate change. In J.M. Guzman, G. Martine, G. McGranaghan, D. Schensul and C. Tacoli (Eds.), *Population dynamics and climate change* (pp. 45–63). New York/London: UNFPA & IIED.

Schremmer, C. and Stead, D. (2009). Restructuring cities for sustainability: A metabolism approach. In M Freire (Ed.), Proceedings of the 5th Urban Research Symposium *Cities and climate change: Responding to an urgent agenda* (pp. 1–20). Washington, DC: World Bank.

SEDAC/CIESIN (n.d.). *Gridded Population of the World and the Global Rural-Urban Mapping Report*. Retrieved from http://sedac.ciesin.columbia.edu/gpw/ on 22 August 2015.

Sethi, M. (2011). Alternative perspective for land planning in today's context: Minimising compensation issues. *Indian Valuer, 43*(4), 443–8.

Singh, S. (2011). Urban governance and the role of citizen participation: A case of small and medium towns of India. *Urban India, 31*(1), 138–53.

Siwach, R. (2011) Capacity building of urban local bodies for urban development: Some emerging areas for NGO intervention. *Urban India, 31*(1), 15–31.

UN (2007). *City planning will determine pace of global warming*. Retrieved from www.un.org/press/en/2007/gaef3190.doc.htm on 5 April 2014.

UN DESA (2011). *Urban population development and the environment wallchart*. New York: United Nations.

UN DESA (2012). *World urbanization prospects: The 2011 revision*. New York: United Nations Population Division, Department of Economic and Social Affairs.

UNEP (2012). *GEO5 report: Measuring progress – environmental goals & gaps*. Nairobi: United Nations Environment Programme.

UNFPA (2007). *State of the World Population 2007: Unleashing the potential of urban growth*. Geneva: UNFPA.

UN-Habitat (2006). *State of the world's cities report 2006/2007*. London: Earthscan.

UN-Habitat (2010). *Climate change strategy 2010–2013*. London/Washington, DC: Earthscan.

UN-Habitat (2011). *Cities and climate change: Global report on human settlements*. London/Washington, DC: Earthscan & UNCHS.

van Vuuren, D.P., Lucas, P. and Hilderink, H. (2007). Downscaling drivers of global environmental change: Enabling use of global SRES scenarios at national and grid levels. *Global Environmental Change, 17*, 114–30. doi:http://dx.doi.org/10.1016/j.gloenvcha.2006.04.004.

Victor, D.G., Zhou, D, Ahmed, E.H.M., Dadhich, P.K., Olivier, J.G.J., Rogner, H.-H., Sheikho, K. and Yamaguchi, M. (2014). Introductory chapter. In *Climate change 2014: Mitigation of climate change*. Contribution of Working Group III to the Fifth Assessment Report of the Intergovernmental Panel on Climate Change [O. Edenhofer, R. Pichs-Madruga, Y. Sokona, E. Farahani, S. Kadner, K. Seyboth, A. Adler, I. Baum, S. Brunner, P. Eickemeier, B. Kriemann, J. Savolainen, S. Schlömer, C. von Stechow, T. Zwickel and J.C. Minx (Eds.)]. Cambridge/New York: Cambridge University Press.

World Bank (2010). *Cities and climate change: An urgent agenda*, Washington, DC: IRDC.

World Resources Institute (2016) *World greenhouse gas emmisions: 2005*. Retrieved from www.wri.org/resources/charts-graphs/world-greenhouse-gas-emissions-2005 on 2 June 2016.

# 2 Interrelation between development, urbanization and carbon footprint

> Global climate change in the 21st century will depend on the interaction of three trajectories – population growth (which is essentially going to be in urban areas of developing countries), economic growth and GHG emissions.
>
> (Martine, 2009, p. 10)

As evident from discussions made in the introductory chapter, post-industrial growth and development activities have led to prolonged accumulation of GHG emissions in the atmosphere for the past 200 or so years. In accordance with the first objective of this research, it is important to understand how this phenomenon unfolds across the world, and how India scores in this. This chapter addresses it with a comprehensive study of the interrelationship between economic development, urbanization and GHG emissions (carbon footprint), by first acknowledging the literary associations between: (1) economic development and GHGs, which is epitomized with North–South debates in the global environmental governance; and (2) urbanization and GHGs, which is a relatively newer area of interest. This is followed by augmenting discourse on the interrelationship between economic development, urbanization and GHGs, with evidence from global and Indian historical data, using time-series analysis. In order to seek clarity on this subject, spatial analytics are applied, which for the first time considers urbanization as an independent parameter and, in accordance with the development status of different countries, recognizes or differentiates these carbon footprints.

## 2.1 Economic development and GHGs

There is a growing body of work that explains how different states of development across the world cause variations in GHG emissions. Highly industrialized, developed countries with higher economic output are the most GHG-emitting, while countries with moderate to low levels of industrialization and economic development have smaller carbon footprints (see Table 2.1 and Figure 2.1).

*Table 2.1* Countries classified as per their economic development and emission status

| Countries classified on economic development status | GDP per capita (constant 2005 US$) | CO₂ emissions (tons per capita) |
| --- | --- | --- |
| High-income | 30,826 | 11.6 |
| Upper-middle-income | 3,931 | 5.4 |
| Middle-income | 2,507 | 3.4 |
| Lower-middle-income | 1,139 | 1.6 |
| Low-income | 406 | 0.3 |

Source: Data classification by the author, data from World Development Indicators

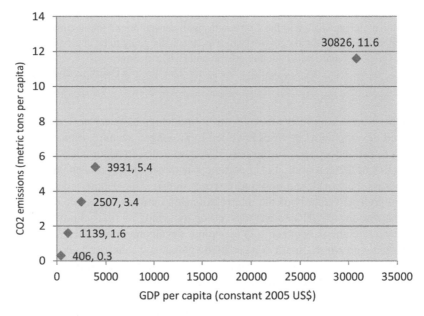

*Figure 2.1* Countries plotted as per their economic development and emission status
Source: Author, data from World Development Indicators

Heil and Wodon (1997) use the Gini index to measure the inequality of per capita emission across countries and the contribution of two income groups (poor and rich countries) to this inequality. They later employ this methodology for analysing future inequality in per capita emissions using business-as-usual projections to the year 2100 (Heil and Wodon, 2000). One of the main conclusions is that both emission and income inequalities have decreased during past decades, later supported by Padilla and Serrano (2006). They further emphasize that "if you belong to the same income group (per capita), emission inequality will be less profound", especially in the low-middle-income group. Duro and Padilla (2006) applied the decomposable Theil index by decomposing the Kaya identity and found that income per capita, or simply put affluence, is the main driver of emission inequality, although differences in energy intensity and carbon intensity of energy were also relevant. In addition, differences in per capita incomes between the richest and poorest persons, in both rich and poor countries, are bigger than the differences in per capita incomes between countries. But if all countries in fact had completely egalitarian income distributions, there would still be huge emission differences based on the gaps in national incomes (Milanovic, 2005). The research in this area underscores that global policies seeking equitable use of atmospheric space would need to reduce deeply rooted income inequalities between the national economies of the Global North, South and within, in addition to expanding technical improvements in the energy intensity and carbon intensity of energy.

The above findings are reinforced by another set of findings challenging our conventional approach to the development and environmental complex. It is established that, unlike other forms of environmental pollution, global warming could not be explained through the famous environmental Kuznets curve (a hypothesized relationship between environmental quality and economic development: various indicators of environmental degradation tend to get worse as modern economic growth occurs until average income reaches a certain point over the course of development, thereafter improving; represented by an inverted u-shaped curve), and appropriate policies are required to reach a turning point in the relationship between income and emissions (Cantore and Padilla, 2009). Furthermore, Groot (2010) exhibits that standard tools in the measurement of income inequality, such as the Lorenz curve and the Gini index, can successfully be applied to the issues measuring carbon emissions and their abatement policies across countries. They allow policymakers and the general public to grasp at a single glance the impact of conventional and sophisticated carbon allocation issues. The principle settled in the UNFCCC is the "common but differentiated responsibilities", or CBDR, according to which more developed countries have greater responsibilities to take action on climate change. Traditionally, the most visible distinction of the CBDR cuts the globe across the lines of developed countries (North) and the developing countries (South), commonly known as the North–South (NS) divide. This also divides the countries legally (e.g., Annex I and Non-Annex I) and the negotiation dynamics (G77+China versus OECD countries).

## 2.2 North–South divide in global environmental governance

The vast disparity in economic status of different societies/countries has led to the so-called NS divide in global environmental governance, particularly the climate debates. The North signifies the highly developed and industrialized countries, while the South refers to the developing ones. Undeniably, a part of the conflict has a strong factual basis. The North has consumed more than its fair share of the Earth's atmospheric space. The cumulative emissions of the North form the significant majority of all historical emissions (Climate Justice, n.d.; Miguez and Domingos, 2002), while per capita emissions of the North have been historically 10 times greater than the South (Kartha *et al.*, 2012), and even now (8.8 t/capita) is over twice that of the South (4.2 t/capita). Second, it is an accepted fact that the cause of global warming was not global in the first place, but rooted in human activities taking place at local and national levels, particularly the industrialization process in the North (Kato, 2001; UNFCCC, 2014).

Third, when it comes to climate vulnerabilities, low-income countries are the most severely affected. Some of the well-rehearsed poverty-related climate effects include an increase in the frequency and severity of extreme climate events, reduced crop yield, which gives rise to food insecurity, lower incomes and scant economic growth, the displacement of the poor from coastal areas, and exposure to new health risks (Adger *et al.*, 2005; Richards, 2003). Accordingly, for reasons of limited infrastructure and wealth, developing countries have the least capacity to respond to this challenge (Bulkeley and Newell, 2009; Climate Justice, n.d.; IPCC, 2007). But beyond the facts, most of the NS disparity is because of diverging interests and motives. The North has its share of concerns for large-scale emission cuts proposed, coupled with financial and technological investments in climate mobilization, including massive support to the South, fear of a rising Asia, and its stubborn insistence that the South is both unwilling and unable to restrain its own emissions (Kartha *et al.*, 2012). Beyond paying lip service to commitment, responding to this challenge will require inherent change in its consumption patterns, lifestyle and business practices. Some of the differences in factual positions, concerns, interests and motives between the Global North and South are enumerated in Table 2.2.

On the contrary, the South's concerns are as numerous and multifarious as its composition. To begin with, with 300 years of industrialization, they consider developed countries as the primary culprits for today's climate problem (Climate Justice, n.d.; Gupta *et al.*, 2000). Second, many of the concerns are rooted in systematic discrimination of the South in the past (Mahabub-ul-Haq, 1976, p. 167) and the North's protracted history of self-interested and bad-faith negotiations in all sorts of other multilateral regimes such as trade. Third, the North has repeatedly failed to meet UNFCCC and Kyoto commitments to provide technological and financial support for mitigation and action, holding the South a hostage to its newly made commitments while continuing to dodge its own (Kartha *et al.*, 2012). The South thus vehemently supports the idea of recognizing historical emissions. It seeks equality in access to global commons, which is epitomized with a possible agreement to grant equal per capita emission right of cumulative carbon space. Any divergence from the South's demand is seen by itself as 'locking-in injustice', where the poor person's share would be significantly reduced. The 'carbon market' is termed as a 'land-grab' that would allow the North to purchase more of the budget from the poor at low prices, encouraging a paradoxical crisis of "Polluter profits and poor pays?" (Climate Justice, n.d.).

In addition to the above, the South has its priorities in physical and socio-economic development. What has been viewed as the South's aspiration of right to development (Gupta *et al.*, 2000; Kartha *et al.*, 2012) is in fact ingrained in its value of existence and sustenance. For small island states, low elevation and mountainous countries, climate-induced changes in their immediate surroundings

*Table 2.2* Diverging North–South positions

| S. no. | North | South |
|---|---|---|
| | *Factual differences* | |
| 1 | Higher cumulative, per capita and historic emissions | Lower cumulative, per capita and historic emissions |
| 2 | Increasing rate of emissions from the South | Large-scale vulnerability to and threat from catastrophic events |
| 3 | Less vulnerable to catastrophic events | Issue of survival for people living on small island states and lower elevations |
| 4 | Higher technical, financial and institutional capacities to respond to mitigation, adaptation and catastrophic challenge | Limited capacities to respond to mitigation and adaptation and catastrophic challenge |
| | *Concerns, interests and motives* | |
| 1 | Concern of large-scale emission cuts, forcing radical shift in lifestyle, behavioral and business patterns | Failure of the North to meet Kyoto commitments |
| 2 | Large-scale financial and technological investments at home | Subjugation of the South, systematic discrimination in the past |
| 3 | Massive support and financial assistance to the South | Impairment to right to economic development |
| 4 | Fear of rising Asia and restructuring of world order | Threat to livelihood security, basic needs and energy access – issue of survival |
| 5 | | Issue of justice – to attain compensation from threats and impacts not caused by their own action |
| 6 | | Means to restructure the world order and having greater role in decision-making |

is not a matter of perception or choice for selecting a development paradigm, but a question of life and death. Leaving a developing nation's economic growth aside, an average citizen of the South struggles to sufficiently avail him/herself of affordable energy, sustain his/her livelihood and raise material living standards. As numerous studies and reports underpin this over and over again, access to energy services is fundamental to the fulfilment of any development goals (Purkayastha, 2010; UNDP/WHO, 2009). Hence, the bottom line is that energy access is central to the issue of the South's existence/survival first, and then the fulfilment of its development goal for the burgeoning population.

In addition to the South's priorities of economic development, sustenance and corrective/restorative justice, there are political overtones to this issue, which could be better understood while looking beyond the economic connotations of NS (Atapattu, 2008; Najam, 2005). The South Commission (1990, p. 1) defines the 'South' as not only characterized by economic weakness, but political dependence on the 'North', which makes them "vulnerable to external factors and lacks functional sovereignty", thereby it undermines their own control over their destinies. Najam (2005, p. 113) argues that the 'South' is a definition of exclusion, and it includes those states that have been overlooked in international decision-making. They view themselves as "existing on the periphery". As such, it is being widely acknowledged that the climate debate is one space where South states demand not only economic justice, but have found an opportunity to alter this structural inequality and play a fundamental role in global decision-making while balancing the world order (Najam, 2005; Roberts & Parks, 2007). Meanwhile, what has been largely understated is that the North also treats the situation as a battleground, and has difficulty accepting its declining prominence in the modern world order. As such, both the North and South fundamentally view the climate negotiation as a leverage point, which inherently becomes a self-evolving course seeking incremental justice (and corrective justice for the South), rather than merely a distributive one. Over time, the system or the regime itself tends to become a vehicle for justice.

This has always raised a fundamental question on the role of states in brokering a truly equitable and acceptable solution. There is growing understanding to approach global climate politics beyond the conventional prism of the state, which is considered to be a statist view of international relations (Eckl and Weber, 2007). It will be more fruitful to view the issue on "intra and transnational social and economic divisions" (Newell, 2005, p. 70), other actors, institutions, corporate houses, national and transnational movements, non-governmental organizations (NGOs) and international non-governmental organizations (INGOs), and others, though there are equally strong arguments for bringing the state back, for its capability to police the perpetrators and bring about the legislations (Pradhan, 2013). Thus, there are several limitations in viewing the current climate impasse through the global NS divide, which is reported to be diluting, both in terms of income inequalities, as demonstrated above in section 2.1, and also mitigation efforts, where the South, particularly led by China and India, has shown growing aggregate expenditures in carbon mitigation (Wheeler, 2011).

With so many differentiations and transformations in emission profiles of several states, complicated by their vested motives – both apparent and inherent – it becomes imperative to discern the truly marginalized and under-represented entities in climate debate. At the international level, it is the least developed nations, mountain states and Alliance of Small Island States (AOSIS). At the economic level, it is the poor half of the world's population living marginally with less than $2.50 a day or less (Chakravarty *et al.*, 2009; Shah, 2013). As such, a more nuanced approach to understand carbon footprints, which looks beyond the NS differentiation and internalizes subnational groups, becomes vital. This could be instrumental in actualizing a breakthrough and leading to climate action on the ground, because it is the local groups that are most closely affected by climate change and associated delays from inaction.

But in the last decade or so, the climate science about greenhouse gas stabilization is rapidly advancing, and so is the emission contribution from the developing world. From the Paris Agreement 2016, a growing consensus has now emerged in favour of having globally low stabilization targets that cannot be achieved without the active participation of developing countries, which today emit about half of global $CO_2$ emissions, and whose future emissions increase faster than the emissions of industrialized countries under "business as usual" scenarios (den Elzen and Hohne, 2008). Also, the later the international community commits to their respective national atmospheric targets, the more difficult and costly it will be to achieve them (Stern, 2006). Recent research also informs us of a greater degree of correlation between cumulative emissions and global warming (Allen *et al.*, 2009; Meinshausen *et al.*, 2009), thereby corroborating the historic responsibilities of developed countries to climate change. On a per capita basis, some proposals such as climate debt quantify this emission divide between developed and developing countries as 10:1 (UNFCCC, 2009). Another study compares the actual versus the fair share of (access to) carbon space (Purkayastha, 2010) and concludes that while Annex I countries are entitled to a fair share of 19%, by 2009 they had utilized an actual share of 74%.

Moreover, if in spite of ambitious commitments developed countries actually cut emissions slowly, less carbon space will be left for the developing countries, where they will: (a) incur higher costs per unit of energy to lower emissions; and (b) have to limit energy consumption. It therefore becomes crucial for the developed countries to cut immediately, and make deep cuts. Developing countries are encouraged to grow their way out of economic and energy poverty, but without growing their emissions, and to do so in an increasingly hostile climate. In addition to this traditional NS distinction, there is another less explored line that has divided the world, in terms of the "common but differentiated responsibilities": *the urban–rural divide.*

## 2.3 Urbanization and GHG emissions

More than half of the world's population has become urbanized for the first time in human history (UN DESA, 2012), with huge implications on consumption of resources, climate change and in terms of owing responsibilities for mitigating the possible impacts. At the aggregate level, almost all population growth is occurring in cities: population growth issues are thus primarily urban issues. The number of urban dwellers will continue to rise quickly, reaching almost 6.2 billion people in 2050 (Martine, 2009). It is believed that most of this growth is taking place in urban areas of low- and middle-income nations, and this is likely to continue (UN, 2010). Thus, a concern for how the growth in the world's population influences GHG emissions is largely a concern for how the growth in the urban population in low- and middle-income countries influences GHG emissions (Satterthwaite, 2009, p. 45).

A study comparing different contributions of nations to population growth and $CO_2$ emissions from 1980 to 2005, when they are classified by average per capita income levels (low-income, lower-middle-income, upper-middle-income and high-income nations), reveals some interesting facts. Nations classified as low-income in 2005 contributed far more to global population growth (52.1%) between 1950 and 2005 than they did to $CO_2$ emissions growth (12.8%). On the contrary, nations classified as high-income in 2005 accounted for far more $CO_2$ emissions growth (29.1%) than population growth (7.2%) (Satterthwaite, 2009, p. 53).

These facts are important because about 87% of the world's future urban growth will be concentrated in developing countries, especially in Africa and Asia. While currently these two continents lag far behind others in terms of urbanization levels, the present and future growth in absolute numbers of urban people in these regions is massive and unprecedented. This transformation will have enormous implications for climate change, given the increasing concentration and magnitude of economic production in urban localities, as well as the higher living standards that urbanites would enjoy in comparison to rural populations. Urban concentration will also be critical for mitigation and adaptation efforts in view of the greater vulnerability of urban populations to some of the more hazardous consequences of global climate change (Martine, 2009, pp. 16–17).

Interestingly, one advantage that potentially benefits developing countries is that much of their urban growth is still to come, giving them the opportunity to make more sustainable use of space at lesser human and financial cost. Taking advantage of this opportunity, however, will require a radical change in the anti-urbanization stance taken by many leaders and policymakers in developing countries, who still try to impede or slow urban growth rather than prepare ahead for it (Martine, 2009, p. 22). Asia alone added 750 million urban dwellers in 20 years (1990–2010), more than the population of Western Europe and the US put together, and it is expected to add another 1 billion in the next 20 years (ADB, 2012). Africa is expected to urbanize rapidly in the next 20 years, adding another 500 million to its cities until 2040 (UN-Habitat *et al.*, 2014). Hence, the growing urban–rural divide poses a formidable challenge to global change and its governance.

It is reasoned that population growth can only be a significant contributor to GHG emissions if the people that make up this growth enjoy levels of consumption that cause significant levels of GHG emissions per person or, from the production perspective, live in nations with a rapid increase in GHG-generating production (Satterthwaite, 2009, p. 54). While across the globe there is an unparalleled demand for fossil-carbon to fuel national economies, it is their urban centres that act as the guzzling engines of energy and carbon. Some accounts are strongly associated with production and consumption of energy within cities. They indicate that urban areas produce more than 70% of the global greenhouse gases (IEA, 2012; Stern, 2006) and consume 60–80% of final energy use globally (GEA, 2012). The issue is of a serious concern for urban areas located in the developing world, because as these countries urbanize, the contributions of carbon emissions and other GHGs from cities will start becoming disproportionately high in comparison to their population share (Satterthwaite, 2009), for example in China. While China attained 50% urbanization in 2011, 40% of the country's $CO_2$ emissions came from the largest 35 cities, though their population was only 18% of the total (Dhakal, 2009).

Developing countries are desperately trying to emulate the lifestyles and consumption practices of industrialized societies. Although at the aggregate level they

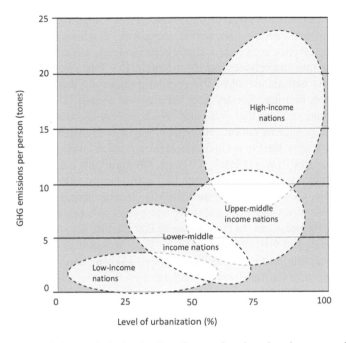

*Figure 2.2* Level of urbanization of countries plotted against per capita greenhouse
gas emissions ($CO_2$e) for 2005

still have a long way to go, they are already starting to make their own massive
impact on global climate change (GCC). Solving this conundrum will require redefining not only 'development', but also the strongly material content of modern-day
'happiness' (Martine, 2009, p. 27). Hence, urbanization can be viewed as one of
the most serious 'problems' causing climate change, in that, in general, the more
urbanized a nation, the higher the GHG emissions per person (Satterthwaite, 2009,
pp. 55–6). Figure 2.2 tenders a generalized representation of the per capita GHG
emissions for 2005 (in t $CO_2$e, based on the production perspective) of different
types of nations (low-income, lower-middle-income, upper-middle-income and
higher-income), plotted against their level of urbanization. The visual divulges
some remarkable facts. Most low-income nations have less than half of their population in urban areas, and many have less than a quarter. Many have per capita
GHG emissions below 0.2 t/year and very few above 2.5 t/year. But note that all of
the upper-middle and high-income nations, and many of the lower-middle-income
nations, had GHG emissions per person above the 'fair share' level, with the US and
Canada having more than 10 times that level. Low-income nations that have little
or no economic growth probably have little or no growth in GHGs in their urban
areas, just as they generally have little or no increase in their urbanization levels
(Potts, 2009).

But for low- and middle-income nations that become wealthier (which also
means becoming more urbanized), the location of consumers and the changes in

their consumption behaviour become increasingly important contributors to GHG emissions (Satterthwaite, 2009). It is further observed that in high-income nations, there are also many manufacturing and service enterprises that are located in rural areas. But here, the division between rural and urban in terms of employment and access to infrastructure and services has disappeared. In effect, virtually all rural areas are 'urban' in that almost all of the population does not work in primary activities (including farming, forestry and fishing) and almost all enjoy levels of infrastructure and services that were previously only associated with urban locales. Thus, in high-income nations, there can be a large increase in per capita GHG emissions and very little or no increase in urbanization levels.

Another notable study (World Bank, 2010, p. 18) highlights the above nexus by studying nine major world economies: the US, the 50 largest cities, C40 Cities, China, Japan, India, Germany, Russia and the UK. It underscores that it is not surprising that rich nations and cities use more energy than poor cities, and therefore emit more GHG emissions. In fact, it asserts that the link between economic growth, urbanization and GHG emissions is by now accepted as a basis from which to start discussing future alternatives. This also substantiates why mitigation is equally important for transforming societies.

## 2.4 Economic development, urbanization and GHG emissions

The interest is increasing in the carbon footprint of cities and geopolitical regions (Peters, 2010). Based on recent research in this area (Dhakal, 2009; Parshall *et al.*, 2009; WEO, 2008), it could be suggested that the present and future levels of urbanization, particularly the rapid urbanization of developing countries, have clear linkages to global GHGs (Dhakal, 2010). It has been further argued that global climate change in the 21st century will depend on the interaction of three trajectories – population growth (which is essentially going to be in urban areas of developing countries, particularly in Asia and Africa, as discussed in section 2.3), economic growth and GHG emissions (Martine, 2009; World Bank, 2010, p. 18).

Global data sets of these three parameters (see Table 2.3) reveal that while more developed regions are 75% urbanized and growing modestly at an annual growth rate of 0.7% (2005–2010), less developed regions are 45% urbanized and expanding at 2.4%, and the least developed countries, though 29% urbanized, are fast multiplying at 4%. High rates of urbanization, considered an outcome of industrialization (just like GHG emissions) are thought to be particularly affecting low-income and middle-income nations. Although the average global emission is 5.8 t $CO_2$/capita, it also varies across more developed, less developed and the least developed regions at 12.0, 3.4 and 0.3 t $CO_2$/capita, respectively. The time-series analysis for global urbanization, economic development and GHG emissions for the last five decades (1960–2010), generated for the first time ever, is shown in Figure 2.3. In this regard, India, with its sheer urban population mass of about 367 million, is urbanizing at an exceptional rate of 2.3% per annum. For the first time since India's independence, the absolute increase in population is more in urban areas than in rural areas (Census of India, 2011), posing a grim environmental challenge.

*Table 2.3* State of urbanization, economy and emissions across major world regions

| Regions | Total population (in thousands) 2010 | Urban population (in thousands) 2010 | Urban population (% of total population) 2010 | Annual rate of urbanization (%) 2005–2010 | GDP/capita at PPP 2005 constant international dollars 2009 | Carbon emissions (t/capita) 2007 |
|---|---|---|---|---|---|---|
| World | 6,895,889 | 3,479,867 | 50 | 1.9 | 9,547 | 5.8 |
| More developed regions | 1,235,900 | 928,853 | 75 | 0.7 | 28,670 | 12.0 |
| Less developed regions | 5,659,989 | 2,551,304 | 45 | 2.4 | 5,218 | 3.4 |
| Least developed countries | 32,330 | 242,769 | 29 | 4.0 | 1252 | 0.3 |
| India | 1,224,614 | 367,507 | 30 | 2.3 | 2,993 | 1.4 |

Source: Data classification by the author, data from UN DESA (2012), later revised 2014

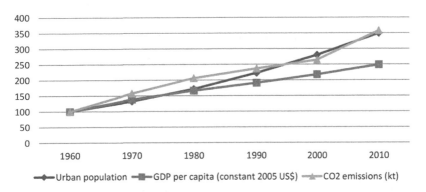

*Figure 2.3* Time-series analysis of world's urbanization, economic development and GHG emissions (index 1960–2010; urban population in thousands)

Source: Data from UN DESA (2012), later revised 2014

## 2.5 Spatial framework to analyse carbon footprints

At present, there is no single benchmark to evaluate cross-sectoral growth in urbanization, economic development and GHG emissions of countries and regions. As of now, GHGs have been studied separately as a dependent variable of either economic growth or urbanization on a single time frame. Second, studies have given overemphasis to 'income level' or GDP of countries as an indicator, which merely represents one of the development indices and rather fuels debates on NS inequity. Any new frame of analysis needs to be based upon a comprehensive and interdependent understanding of: (1) climate governance; (2) principles of ethics, fairness and justice; and (3) carbon access and allocation. For a complete theoretical study, see Sethi (2015). The proposed spatial framework intends to further analyse the ongoing phenomenon of global change by studying the dynamic relationship between development, urbanization and GHG emissions.

This research applies spatial analysis to add a few additional dimensions to this dynamic relationship. It puts forward a framework that considers GHG emissions of a country as a consequence of, first, its spatial circumstances and, second, its development status over a long-term dimension of time. The spatial circumstances of a country are understood from its level of urbanization, whether it is rural, urban or urbanizing, and treated as an independent variable on the *x*-axis. In the absence of any universal definition by which all nations could be classified as urban or rural, we consider defining it as per three equal states of urbanization on a ratio scale of 1 to 100 as $0 < x \leq 33$, $34 \leq x \leq 67$ and $x \geq 67$, where $x$ is urban population (in percent) for a country/territory for a particular data year.

The development status of a country is not defined using the convention of GDP or average income level, but from the UN classification of countries, which is more generic and inclusive (i.e. more developed regions (MDRs) and less developed regions (LDRs)). Data for the countries that are considered as the most underdeveloped and marginalized (i.e. least developed countries (LDC)) by the United Nations is reclassified from the LDRs. This would help to better comprehend their circumstances while excluding leading countries from the LDR group

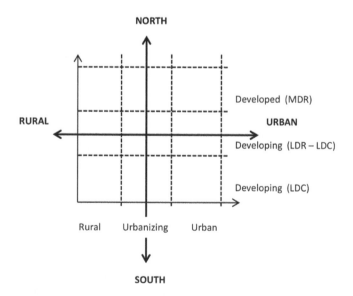

*Figure 2.4* The analytical framework – a 3×3 spatial development matrix
Source: Sethi (2015)

such as China, India, South Africa, Brazil, Indonesia, etc., thus resulting in a 3×3 spatial development matrix, shown in Figure 2.4. This frame is used to analyse GHG emissions over a period of 50 years from 1960 to 2010.

## 2.6 Data and research method

This research utilizes standard definitions and universally available global data sets on emissions, urbanization and country classifications to disaggregate emissions of over 200 countries/territories according to their development circumstances and rural–urban existence in space at various temporal points across six decades (1960–2010).

A **country/territory** is defined using the UN classification of development regions followed in the *World Population Prospects* and several other international studies. A conscious effort to select such a definition (beyond economic indicators based on GDP) has its basis in the literature findings that suggest that the NS divide is not merely an economic divide, but a geopolitical one. Accordingly, MDRs, or the developed countries, comprise Europe, North America, Australia/New Zealand and Japan, and LDRs, or the developing countries, comprise all regions of Africa, Asia (excluding Japan), Latin America and the Caribbean, plus Melanesia, Micronesia and Polynesia. The LDC, as designated by the United Nations General Assembly in 2010, comprises 49 countries, including 33 in Africa, 10 in Asia, one in Latin America and the Caribbean, and five in Oceania. These countries are: Afghanistan, Angola, Bangladesh, Benin, Bhutan, Burkina Faso, Burundi, Cambodia, the Central African Republic, Chad, Comoros, the Democratic Republic of the Congo, Djibouti, Equatorial Guinea, Eritrea, Ethiopia, the Gambia, Guinea, Guinea-Bissau, Haiti, Kiribati, the Lao People's Democratic Republic, Lesotho, Liberia, Madagascar, Malawi, Maldives, Mali, Mauritania, Mozambique, Myanmar, Nepal, Niger, Rwanda, Samoa, Sao Tome and Principe, Senegal, Sierra Leone, the Solomon Islands, Somalia, Sudan, Timor-Leste, Togo, Tuvalu, Uganda, the United Republic of Tanzania, Vanuatu, Yemen and Zambia.

**Urbanization** is defined as the proportion of the mid-year de facto population (in percentage) living in areas classified as urban according to the criteria used by each country or area. Data on urbanization is obtained from *World Population Prospects: The 2011 Revision* (UN DESA, 2012), later revised in 2014.

**Carbon dioxide emissions (tons per capita)** is defined as the annual volume of emissions stemming from the burning of solid, liquid and gas fuels, gas glaring and selected manufacturing processes, including the production of cement, divided by mid-year population and sourced from Oak Ridge National Laboratory, Carbon Dioxide Information Analysis Center (CDIAC) 2007 data available online with the World Bank's World Development Indicators database at http://data.worldbank.org. Biospheric $CO_2$ (emissions from land use and local sequestrization) and other greenhouse gases and aerosols are excluded because they are not strongly correlated with personal choices and national carbon intensities (Costa *et al.*, 2011).

It is understood that rural areas as default act as key sinks, for their sequsterization potential if considered would invariably suggest lower carbon emission emanating from rural areas. By imputing national emissions to urban and rural areas, embedded carbon in exports and imports is excluded, a component that is relevant for countries with large shares of trade in their economy. Historical responsibility prior to 1960 is not considered due to lack of reliable data on emissions and urbanization for developing countries.

**Energy consumption.** In addition to the above, analysis for the current state considers carbon emissions along with energy consumption trends on the 3×3 spatial development matrix. Data for energy consumption (kg of oil equivalent per capita) is the annual consumption of primary energy – primary electricity, crude oil, natural gas, solid fuels (coal, lignite and other derived fuels) and combustible renewables (e.g. wood), divided by mid-year population. It is equal to indigenous production plus imports and stock changes, minus exports and fuels supplied to

*Table 2.4* Synchronization of data from UNDESA and the World Bank database

|  | *UN database on country classification and urbanization (UN DESA)* | *World Bank database on emissions (from CDIAC), gross domestic product (World Bank) and energy use (from IEA)* | *Details* |
|---|---|---|---|
| More developed regions | 56 | 53 | Gibralter, Holy See, Saint Pierre and Miquelon |
| Developing regions (excluding LDC) | 124 | 106 | Mayotte, Reunion, Western Sahara, Saint Helena, Occupied Palestinian Territory, Anguilla, British Virgin Islands, Guadeloupe, Martinique, Montserrat, Netherlands Antilles, Falkland Islands (Malvinas), French Guiana, Nauru, Cook Islands, Niue, Tokelau, Wallis and Futuna Islands |
| LDC | 49 | 50 | All 49 samples match + 1 (South Sudan) |
| Total number of countries | 229 | 209 | |
| Additional countries in WB database | Kosovo, Sint Maarten (Dutch part), St. Martin (French part) | | |

Source: Sethi and Puppim de Oliveira (2015)

ships and aircraft engaged in international transport. It has been sourced from the World Development Indicators database, available online at: http://data.world bank.org, originally from the International Energy Agency (WEO, 2008). All the data sets were superimposed in MS Excel 2007, keeping the urbanization data of 229 countries/territories from the UN as the base (see Table 2.4). The LDC group was reclassified from developing regions, and all three groups, MDR, LDR minus LDC, and LDC, were identified in the database. Data for 20 countries/territories were not available in World Bank data set.

Hence, emissions data of 209 countries/territories were disaggregated using standard sort and filter operations, for all three regions (MDR, LDR minus LDC, and LDC) on the urban–rural gradient, for three categories – urban, urbanizing and rural – numerically defined as ($0 < x < 33$, $34 < x < 67$ and $x > 67$, where $x$ is urban population (in percent) for a country/territory for a particular data year for all six decadal points (1960–2010), and tabulated (see Annexure I) to generate: (1) a 3D surface, which shows trends in values across two dimensions (development state and urbanization) and carbon emissions as the third dimension forming a continuous three-dimensional surface; and (2) a contour, which resembles a surface chart viewed from above. In both the case, shades represent ranges of values (carbon dioxide emissions in tons per capita). For a complete understanding of this research method, see Sethi and Puppim de Oliveira (2015).

## 2.7 Indicator test: urbanization vs. GDP

The analysis begins by responding to what some argue as to whether 'urbanization' qualifies to be an appropriate indicator for $CO_2$ emission trends. For this, its association with emissions is tested against the most common parameter of economy used in most of the schemes i.e. gross domestic product (GDP) employing GDP per capita at PPP, 2005 constant international dollars as the indicator. Co-relation analysis is used to test five equation types – linear, polynomial, logarithmic, exponential and power – for their Karl Pearson coefficient in which the polynomial equations coincidently generated the highest values across all type of data sets. For global data of 209 countries/territories, economy demonstrates a positive quadratic association with respect to per capita emissions ($Y = -0.0003x^2 + 0.0946x$; $R^2 = 0.0889$). For other regions, the MDR, LDR minus LDC, and LDC all show differential results, with LDR minus LDC showing the strongest coefficient of co-relation (right-hand graphs in Figure 2.5). Meanwhile, globally, urbanization has a positive polynomial association with per capita emissions ($Y = 0.0018x^2 - 0.0774x + 2.687$; $R^2 = 0.2258$).

All subgroups (i.e. MDR, LDR minus LDC, and LDC) exhibit positive association, with LDC showing the strongest coefficient of co-relation (left-hand graphs in Figure 2.5). Employing universal data, these findings reinforce contemporary studies on urbanization emissions that used sampled data sets such as Jorgenson and Clark (2010) ($N = 86$ developed countries and LDC from 1960 to 2005) and Poumanyvong and Kaneko (2010) ($N = 99$ countries of all income

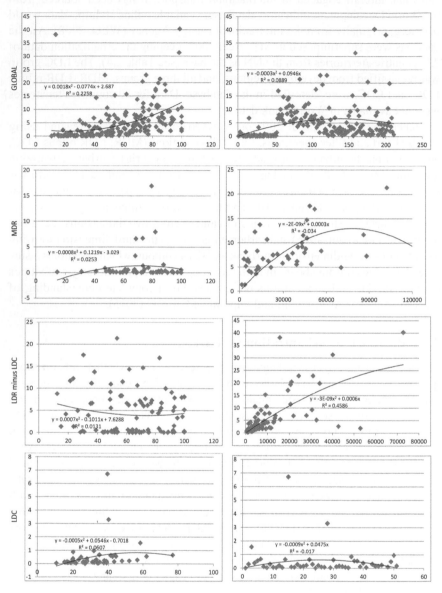

*Figure 2.5* Comparative results from co-relation analysis of (1) urbanization and carbon
dioxide emissions ($X$ = percentage of urban population, $Y$ = carbon dioxide
emissions in metric tons/capita), shown on left-hand side, with (2) state of
economy and carbon dioxide emissions ($X$ = with GDP per capita at PPP, 2005
constant international dollars, with scale adjusted as the indicator, $Y$ = carbon
dioxide emissions metric tons/capita) for global data of 209 countries/territories,
MDR, LDR minus LDC, LDC in 2010, shown on right-hand side

Source: Adapted from Sethi and Puppim de Oliveira (2015)

groups from 1975 to 2005). It supports urban political economy approaches that indicate that cities, being centres of large populations and economic activities, heavily contribute to the burning of fossil fuels and anthropogenic emissions. Although numerically indicators of economy (GDP/capita) and urbanization, both have limited association with increasing $CO_2$ emissions, but it is notable, all things being equal, in comparison to GDP, the urbanization level of countries worldwide exhibit a much greater $R$-square value with $CO_2$ emissions (i.e. 0.2258>>0.0889). Based on normative and empirical knowledge of associating GDP with country emissions so far, by logic of relative judgement (of higher $R$-square values), makes urbanization a sufficiently qualifying indicator for further experimental investigation. It could be argued that inequality in the distribution of emissions between developed, developing and LDC countries could be reasonably explained by differences in the levels of their urbanization than their economic inequalities (in terms of GDP/capita), as had been the norm. While in the past economic inequalities were used to expound emission inequalities, this assessment suggests that, globally, the state of urbanization of societies is indeed an equally good, or rather numerically better, indicator to analyse their carbon emissions.

## 2.8 Discussion of results: unfolding of the phenomenon

The results generated in the form of two visualizations in Figure 2.6 – (1) a 3D surface; and (2) a contour – could be classified into three broad findings, as discussed below. For more insights of results and implications on international climate governance, see Sethi and Puppim de Oliveira (2015).

### 2.8.1 Diluting of the NS emission disparity

In 1960, high emissions were concentrated in the Global North (more developed regions) with 6.09 t/capita. Less developed regions, excluding LDC, had average emissions of 1.47 t/capita, while LDC emissions were 0.11 t/capita. In 2010, the emissions for these regions became 8.18, 5.51 and 0.49 t/capita, respectively. Although the average NS disparity has since changed numerically from 5.98 to 7.69 t/capita, the NS emission differential has reduced substantially from over 55 to about 17 in 2010. The difference between the two most extreme constituencies (i.e. urban North and rural South) has reduced from 10.11 t/capita in 1960 to 8.85 t/capita in 2010. The basic emissions of the rural South have hardly grown during this half-century, from 0.1 to 0.22 t/capita. This reinforces the prevailing knowledge on diminishing income and emission inequalities across the world regions (Padilla and Serrano, 2006), with credible evidence of spatial attributes. The results also point to the fact that, contrary to the traditional view, categorizing past emissions across strict NS lines could be misleading. Though, on average, emissions from developed countries have always exceeded developing regions, there are a lot of exceptions to this general principle when emissions are

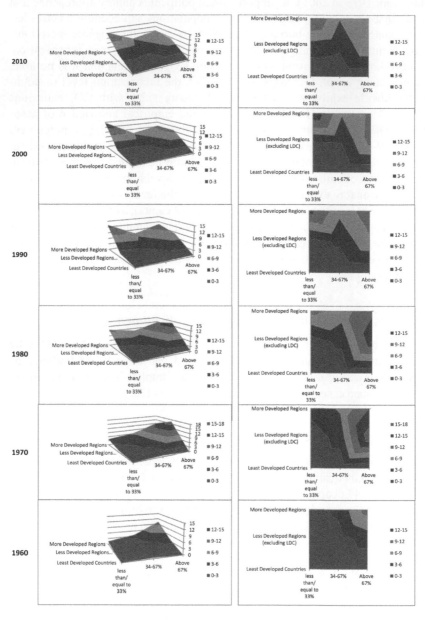

*Figure 2.6* (a) 3D surface, which shows trends in values across two dimensions (development state and urbanization, in this case) and carbon emissions in metric tons/capita as the third dimension forming a continuous three-dimensional surface; and (b) contour, which resembles a surface chart that is viewed from above (in both cases, shades represent range of values; carbon dioxide emissions in metric tons per capita)

Source: Adapted from Sethi and Puppim de Oliveira (2015)

differentiated from the spatial or urban–rural (UR) perspective of their origin. This could be explained with the help of the following findings:

(a) *Differentiated emissions from the North in 1960–1970:* Until the 1960s, and even until the 1970s, high emissions of the order of 10.21 t/capita in the Global North emanated particularly from 14 highly urbanized and prosperous countries only, followed by 29 urbanizing countries with an average of 3.23 t/capita, while there were still about nine countries emitting 1.48 t/capita, which were similar to average emissions in less developing countries, excluding LDC.
(b) *Emissions from urbanized LDR exceed urbanizing MDR:* From 1960 to 2010, the emissions from highly urbanized countries (above 67%) of less developing regions have been greater than urbanizing (34–67%) nations of even developed regions. Their comparative per capita emissions in t/capita are (4.13, 3.73), (16.69, 6.39), (13.84, 7.43), (8.30, 6.90), (9.66, 6.37) and (8.30, 6.21), respectively, for six decadal points from 1960 to 2010.

This reflects that for higher emissions, one's location or spatial circumstances of being in a highly urbanized, urbanizing or rural society could significantly influence per capita emissions rather than national economic or development circumstances, at least while excluding LDC from the argument.

### 2.8.2 Sustained UR emission disparity

In 1960, there were 25 countries with high levels of urbanization (above 67%) with an average emission of 7.17 t/capita, and 74 countries with low levels of urbanization (less than or equal to 33%), with an average emission of 0.28 t/capita. The UR emission differential was 25.60. In 2010, countries with high levels of urbanization (above 67%) rose to 74, with an average emission of 8.54 t/capita. Meanwhile, countries with low levels of urbanization (less than or equal to 33%) reduced to 38, with an average emission of 1.76 t/capita, thus UR emission differential reduced to 4.85. Numerically, this disparity has remained almost constant, from 6.89 t/capita (1960) to 6.78 t/ capita (2010).

In the past, the greatest ever carbon inequalities (of 15.68t/capita and 12.40t/capita) between any two spatial units has been recorded for UR in 1970 and 1980, respectively, in the case of the LDR *minus* LDC group. Analysis of emissions across all vertical and horizontal units in the 3×3 matrix indicates that UR disparity is thriving and has found enough ground in the North. Within the urbanized societies, the emission difference across NS (MDR and LDC) has marginally reduced from 10.21 to 8.43 t/capita during 1960–2010. This reinstates that, irrespective of their geopolitical or economic situation, countries with higher levels of urbanization or cities in general are coming together, becoming even in their per capita carbon throughput.

On the contrary, the emission gap has grown significantly between the rural countries of NS during this period. In 1960 the NS emission differential in rural societies across the globe was 14.8, which in 2010 has expanded to 56.6. While the average emissions of rural LDC such as Liechtenstein, the Channel Islands and the Faroe Islands have increased marginally from 0.1 to 0.22 t/capita, those of rural developed countries have multiplied enormously from 1.48 to 12.45 t/capita. Since 1990, per capita emissions from rural countries in the North have started peaking. This is evident from data and estimated emission figures from small countries with rural populations. The trend started with their +2.36 t/capita emissions above the urbanized societies in 1990, peaking to +5.22 t/capita in 2000, which has relatively stabilized to +3.38 t/capita in 2010. Its causation needs further exploration, whether this is due to their geographical situation or whether it is increasingly becoming energy-intensive to sustain their rural lifestyles and economies. On the contrary, in the LDR minus LDC group and LDC, the per capita emissions in urban societies still exceed rural ones by 3.89 t/capita and 0.42 t/capita, respectively, in 2010, indicating how cities in the developing world outweigh their rural counterparts.

### 2.8.3 Other emerging patterns

**(a) Shifting of polarity.** Data sets of 1970 and 1980 indicate a sudden shift of high per capita emissions towards highly urbanized societies of the LDR minus LDC group (16.69 t/capita and 13.84 t/capita, respectively) although countries from MDR such as the US, Canada, Belgium, Denmark, Australia, the UK, Sweden and the Netherlands also had emissions above 10 t/capita. The high levels in the LDR minus LDC group was not due to a rapid rise in fossil-carbon use or energy consumption within the society, but because of the rise in fossil fuel generation in some newly oil-producing countries such as the UAE, Qatar and Brunei.

**(b) Worldwide stabilization of emissions in the recent past.** Over the last half-century, per capita emissions have grown throughout the world (see Table 2.5): MDR (6.1%), LDR minus LDC (30.2%) and LDC (34.8%). But in the recent past, there is a favourable point of departure from earlier trends. In the last three data sets (1990, 2000, 2010), the North has apparently reached a plateau of high emissions at an average of 8 t/capita and above. This consolidation is throughout the urbanized, urbanizing and rural spatial units of MDR. Similarly, high-emission countries in less developed regions (Qatar 69.22, the UAE 65.85,

*Table 2.5* Cumulative decadal growth rate of emissions (1960–2010) across different spatial units in major world regions

|  | 67% and above | 34–67% | Less than or equal to 33% | Average |
|---|---|---|---|---|
| More developed regions (MDR) | −2.3 | 10.7 | 53.1 | 6.1 |
| Less developed regions (LDR) excluding least developed countries (LDC) | 15.0 | 15.8 | 62.4 | 30.2 |
| Least developed countries (LDC) | NIL | 9.6 | 17.1 | 34.8 |

Source: Sethi and Puppim de Oliveira (2015)

Brunei 63.29, Kuwait 33.37, New Caledonia 22.17, Libya 15.57, Bahamas 15.20, all values in t/capita), incidentally the most urbanized, also stabilized their emissions in the range of 6–40 t/capita by 2010.

**(c) Early incidence to higher emission for developing countries.** On a pathway to urbanization, physical and socio-economic development, the incidence to moderately higher emissions (in the range of 3–6 t/capita) for less developing regions is much greater than ever before. In 1960, the occurrence was synonymous with an overall national urbanization of not less than 67%. Surprisingly, in recent decades, it is evident that nations could experience similar levels of emissions quite early during moderate urbanization rates (34–67%).

**(d) Down South remains carbon poor.** All along, LDC continue to have a very low-carbon emission profile, no matter whether the societies are rural, urbanizing or urban. The emission contour of 3–6 t/capita has still not touched the vast unprivileged populations of LDC residing in Africa and Asia, not even their urbanizing middle class. The emissions within LDC are decoupled from the spatial location of an individual (urban, rural, etc.) and within the most fundamental level of 0–3 t/capita. This could possibly be attributed to their subsistence state of national economies and income levels.

## 2.9 Validation of results

In order to validate the results, carbon emission patterns are analysed with consumption patterns on the ground. This is done by comparison of emissions with energy use trends on a 3×3 spatial development matrix, and supplemented by a summary of case studies on carbon emissions, energy access and energy consumption across the globe. The results are discussed below.

### 2.9.1 Comparison of emissions and energy patterns in the 3×3 matrix

The analysis reveals that energy consumption from fossil fuels is polarized or highly concentrated in the urbanized part of the globe, rather than just the North (see Figure 2.7). Within urban societies of the world, it ranges from 3,500 (for LDC) to 4,500 kg of oil equivalent per capita (for MDR). This is quite a contrast to the emission pattern, which is rather oblivious of the urban South's energy consumption.

Though emissions in the North (particularly urban) have largely reached a plateau around 6–9 t/ capita (a large part of which is due to exporting industrial activities to the developing countries), its energy consumption is still astonishingly high, indicating that emission gains made due to exports in manufacturing industries could be offset by an increase in energy use at home in the near future. Thus, it is important for MDR to particularly regulate consumption of fossil fuels that exacerbate emissions at the domestic level. Owing to the unavailability of country data on energy consumption, high emissions in the rural North could not be corroborated further.

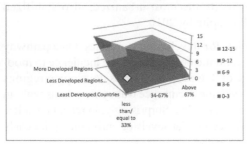
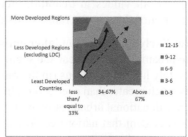
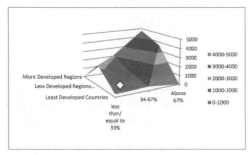
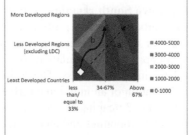

*Figure 2.7* Comparison of carbon emissions in metric tons per capita with energy use trends in kg of oil equivalent per capita for 2010 in the 3×3 spatial development matrix; India's current position is shown as a diamond motif, with a dotted line (a) as the business-as-usual scenario, while a solid line demonstrates the desired low-impact development trajectory

### 2.9.2 Case studies in energy access and consumption

A summary of global case studies in energy access and consumption further substantiates the findings from the 3×3 matrix (Sethi, 2015). On a primary plane, spatial disparity in carbon could be understood from the IEA's World Energy Outlook (WEO) data on energy access (though this may invariably include non-fossil-carbon-derived energy forms too). It captures access to electricity as electrification rates in urban and rural areas in various countries across world regions. It demonstrates that the UR energy gap is typical to developing countries, where urban areas have electrification rates of 90.6%, while rural areas are at 63.2%. Keeping electricity access in urban areas as the yardstick of maximum access within the respective regions as 1.0, the urban–rural energy differential for developing countries is derived as 0.71 (see Figure 2.8). Similarly, for other world regions, such as Latin America, the Middle East, Asia (excluding Japan) and Africa, the urban–rural differential is 0.77, 0.71, 0.77 and 0.35, respectively. This shows relatively high availability and concentration of energy in urban areas, akin to emissions, concluded from the 3×3 matrix. The condition of rural constituencies in South Asia and Africa is at the most disadvantage, as their electrification rates are 59.9% and 25%, respectively, with Sub-Saharan Africa at the bottom with 14.2%.

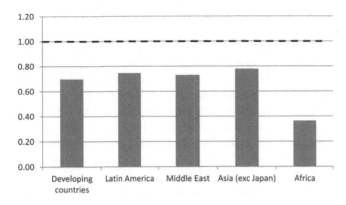

*Figure 2.8* Differential in urban–rural energy consumption (keeping urban as
constant at 1.0) in various world regions

Source: Sethi (2015), data from IEA (2011)

Meanwhile, study of selected cases of relative consumption levels of final
energy use (which is an advanced indicator of fossil-carbon consumption and
carbon emissions in some cases) between urban and rural areas of 22 different
countries spread over all world regions, including developed countries, this time in
the comparison, explains the situation even better (see Figure 2.9). For information
on the sources of this metadata, see Appendix 2. Again, for simplicity of units and
comparison, keeping 1.0 as the urban baseline, it demonstrates a great inter-spatial
divide in urban and rural energy use within their respective national contexts.

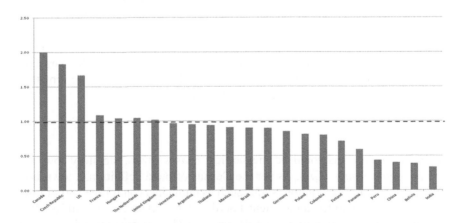

*Figure 2.9* Differential in urban–rural energy consumption (keeping urban consumption
as constant at 1.0) in various countries

Source: Sethi (2015), data compiled from multiple sources (see Annexure II)

Drawing together a novel set of urban energy use metadata, this assessment reveals a huge disparity in urban and rural energy use patterns. In many developing countries, urban dwellers use substantially more final energy per capita than their rural counterparts, as high as two- to threefold in India, Bolivia, China, Peru, etc. This may be primarily because of their much higher average urban incomes. Conversely, in many industrialized countries, the per capita final energy use of rural dwellers is often greater than the urban, by a factor of 1.2 to 2.0, as seen in Canada, the US, France and the Czech Republic. The results validate that in most parts of the world, urban areas in general have the biggest 'carbon footprint' responsible for excessive fossil-carbon use and emissions beyond their national average or fair share, though with high variability in terms of quantity and their NS origins. Meanwhile, in certain cases of exception, it is the rural areas, particularly from the most developing regions in the North, that are consuming more carbon than their equitable share.

## 2.10 Economic development, urbanization and GHG emissions in India

India, like many other developing countries in Asia and Africa, is beaming under newly found economic liberalization, followed by unreeling urbanization, in the last two decades. But there is limited discourse on urbanization and climate change, particularly about how urban areas contribute to GHGs, and none comes to the fore that estimates the interrelationship between economic development, urbanization and emission trajectories.

### 2.10.1 Economic development

Though the total expenditure on urban infrastructure in India is merely 1.59% of the GDP, the contribution of cities to GDP has been steadily rising from 37.7% in 1971 to 52% in 2001 to 63% in 2011 (HPEC, 2011), which is further forecasted to become 75% by 2031. The relationship between per capita incomes and urbanization levels across Indian states is emerging stronger, as evident in Maharashtra, Karnataka, Tamil Nadu, Kerala, etc., indicating that further economic growth in the country will consolidate urbanization trends.

### 2.10.2 Urbanization

In India, urban areas could be megacities, *census towns*, statutory towns or *urban agglomeration* (UA). As per the Census of India definition, a town is an entity that either has a statutory status such as municipal committee/corporation/notified area committee/cantonment board, estate office, etc., or fulfils all of the following three conditions simultaneously: (1) a population of more than 5,000; (2) more than 75% of the male working population is engaged in non-agricultural activities; and (3) density of population is more than 400 persons/square kilometer (Census of India, 2011). There are 4,041 statutory towns and

3,894 *census towns* in India (TCPO, 2012). The statutory towns have increased by 6.37% and the *census towns* substantially by 185%, signifying that a number of rural areas have now attained urban characteristics and are designated as *census towns*. Out of 7,935 towns in India, 468 towns are Class I (population more than 0.1 million), and 53 are million-plus cities. Almost four out of every five Class I towns has a population of 0.1–0.5 million. The average size of towns and cities in India has grown from 33,624 in 1961 to about 61,159 in 2011. It has also been observed that the growth in big metros is stagnating, while the newer and smaller ones are growing faster.

With a total urban population of 377 million in India, urbanization has increased from 27.81 to 31.16% during 2001–2011 (MoUD, 2011). India is adding population of almost that of four Australias put together to its cities every decade. Urban population is increasing at 2.76% of annual exponential growth rate against rural at 1.15%. The absolute increase in population is more in urban areas than in rural areas on account of net rural urban classification and migration (56%), against natural increase (44%). As evident, it took nearly 40 years (1971–2011) for India's urban population to rise by 270 million, but in future it may take half the time to add the same number. According to various estimates, by 2030, India's urban population will be 590 million (Mckinsey, 2010) to 600 million (MoUD, 2011) (i.e. about 40% of the total population at that time) and will break even with the rural population by 2039. A major initiative from the urban perspective was introduced by the national government in 2005 under the JNNURM to invest more than US$1,492 million in 65 cities for upgrading physical infrastructure, urban transport, housing for the urban poor and good governance. The funding was based on a strategic planning document called "The City Development Plan". It is striking that any aspect pertaining to climate change mitigation or adaptation was not a prerequisite for this strategic vision.

### 2.10.3 GHG emissions

The first official assessment made by India under the UNFCCC's National Communication (NATCOM) framework, following IPCC methodology, was using 1994 data, and accounted for 1,228.54 Mt $CO_2$e (MoEF, 2012). Ever since, national GHGs have been increasing at 2.9% per annum, while individual emissions have also nudged from 1.4 to 1.5 t $CO_2$e per capita. The latest assessment of the Indian Network for Climate Change Assessment (INCCA) under the Ministry of Environment and Forests (MoEF) in 2012 captures a sector-oriented reporting based on UNFCCC-mandated inventories. It estimates GHGs from anthropogenic activities in energy, industry, agriculture, waste and LULUCF at 1,727.71 Mt $CO_2$e (MoEF, 2012). Table 2.6 shows the contribution of different sources, activities and economic sectors.

Cities give refuge to functions or activities that consume vast energy supplies, and in the process emit GHG emissions. It has been reported that while average national emissions were 1.4 t $CO_2$e/capita, that of the major cities such as Mumbai, Chennai and Bangalore were assessed to be 1.3 t $CO_2$e/capita (World Bank, 2010).

*Table 2.6* Contribution of GHG emissions from different sectors in India

| S. no. | Sector | Million tons $CO_2e$ 1994 | Million tons $CO_2 e$ 2007 | Sector contribution in total Indian GHG emissions (%) | CAGR (%) |
|---|---|---|---|---|---|
| 1 | Electricity | 355.03 | 719.3 | 37.8 | 5.6 |
| 2 | Transport | 80.28 | 142 | 7.5 | 4.5 |
| 3 | Residential | 78.89 | 137.8 | 7.2 | 4.4 |
| 4 | Other energy | 78.93 | 100.9 | 5.3 | 1.9 |
| 5 | Cement | 60.87 | 129.9 | 6.8 | 6 |
| 6 | Iron and steel | 90.53 | 117.3 | 6.2 | 2 |
| 7 | Other industry | 125.41 | 165.3 | 8.7 | 2.2 |
| 8 | Agriculture | 344.48 | 334.4 | 17.6 | −0.2 |
| 9 | Waste | 23.23 | 57.73 | 3.0 | 7.3 |
| 10 | LULUCF | 14.29 | −177 | | |
| | Total | 1,228.5 | 1,728 | 100 | 2.9 |

Source: India's GHG Emissions Profile (MoEF, 2012)

Meanwhile, climate change finds a narrow reference in the national policy framework (i.e. National Environment Policy, NEP) (MoEF, 2006). Most of the document clarifies India's position in the international climate change debate rather than offering a nationwide integrated approach on the subject. It upholds the principle of common but differentiated responsibilities and respective capabilities of different countries. Pertaining to mitigation, the policy empha-sizes multilateral approaches, rights to equal per capita entitlements of global environmental resources, priority to the right to develop, and encouragement of Indian industry to participate in clean development mechanisms through capacity building. This was followed by the national-level policy on the sub-ject, namely the National Action Plan on Climate Change (NAPCC), adopted by the central government, which identifies eight missions: National Solar Mission, National Mission for Enhanced Energy Efficiency, National Mission on Sustainable Habitat, National Water Mission, National Mission for Sustaining the Himalayan Ecosystem, National Mission for a Green India, National Mission for Sustainable Agriculture and National Mission on Strategic Knowledge for Climate Change (MoEF, 2008). Themes bearing strong potential to influence urban India are the National Missions on Sustainable Habitat, Energy Efficiency, Solar Mission, Green India and Strategic Knowledge for Climate Change. There are two major concerns where the policy confines its outlook. First, it does not follow an integrated view, but a squarely sectoral approach to contain GHG emissions. Second, it is limited to the identification of the institutional and pro-cedural mechanisms that will enable the action plan to function, so at the most is virtually a vision paper.

But it is worth mentioning that neither the NAPCC nor the NEP exam-ines urban emissions separately. Meanwhile, following international practice, the International Council for Local Environmental Initiatives (ICLEI) South Asia reported emissions for selected Indian cities in one of its reports on

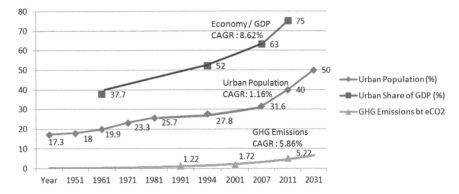

*Figure 2.10* Scenario/time-series analysis of economic development, urbanization and GHG emissions

Source: The time-series analysis generated by the author based on the following data sources:

1  Urban population data from TCPO (2012), projections for 2039 from MoUD (2011)
2  Urban share of GDP data from HPEC (2011), which is based upon CSO and XI Five Year Plan
3  GHG emissions data (1994 and 2007) from MoEF (2012), Indian Network for Climate Change Assessment, MoEF, India and forecasts for 2030/2031 averaged from five models from MoEF (2009)

the subcontinent in 2009. The World Bank (2010) also released "Cities and Climate Change", publishing estimated emissions of seven major metros of India. The study of GHG emissions in 2007 (MoEF, 2010) projects that, on average, national emissions over the next two decades will increase threefold, from 1.72 (2007) to 5.22 t $CO_2$e (2030–2031) while per capita emissions will increase two and half times from 1.5 (2007) to 5.6 t $CO_2$e (2030–2031).

### 2.10.4 Superimposed scenarios

Scenarios of India's urbanization, along with its economy and GHG emissions, derived from time-series analysis, are shown in Figure 2.10. The superimposed scenarios clearly indicate that all the trends are escalating in tandem, calling for an urgency to limit climate change, a situation where cities play a pivotal role in balancing economic development with appropriate climate strategies. In this context, it becomes pertinent to rationally gauge the quantum and nature of emissions generated from urban areas.

## 2.11 Inferences

The research infers that there is a strong interrelationship between the rise in India's GHG emissions with growing urbanization and expanding economy. The results also suggest that India, though it has been traditionally a low-carbon country, is increasingly becoming carbon-intensive. The role of Indian cities in furthering global warming is increasing at a disturbing rate, superseding the growth rates of urban population and economy. The growth is apparently proportionate, or rather a product of both economic growth and the urbanization pattern, thereby indicating incremental emissions owing to the creation of wealth and lifestyle changes. This initiates the need to research urban emissions from a consumption-based perspective.

Without commenting on any causations or co-relations within this association at this stage, while comparing their growth rates, it could be established that a unit growth in the urbanization rate of India is coincident with economy by a factor of 1.43, and with national GHG emissions by a factor of 3.3, the relationship could be denoted with the following equation:

urbanization (100%) <> economy (143%) <> GHG emission (330%)

It indicates a very critical situation for urban development, economic growth and its implications to global warming in the near future. Taking cognizance to forecasts for India's urbanization rate hovering around 40–50% by 2030–2039 (from 32% in 2011), the GHG emissions emanating from cities would expand to 2,160–2,696 Mt $CO_2$e by 2030–2039, if the economy grows constantly at the prevailing rate. Economy being the main driver in this relationship, the comparative growth rate equation could be written alternatively as:

economy (100%) <> urbanization (70%) <> GHG emission (230%)

Accordingly, with varying growth rates in economy, the urbanization and GHG emissions could further be affected. The nexus between economic development – urbanization and GHG emissions could be made favourable in the future by pursing sustainable urbanization. Urban development – planning and governance – has to respond to the call of being low-carbon, green and smart in spite of prospering economically and being globally competitive. This requires further exploration into the causations of GHG emissions at the urban scale itself (which is studied at length in Chapter 4).

India holds a peculiar spot in the 3×3 spatial matrix because it bears a low degree of urbanization (about 31% in 2010) with decent levels of economic development (GDP per capita at PPP of 1,417.06 at 2005 constant international dollars), thereby positioning it in the club of LDR or the developing countries. Considering the clout of India in global geopolitics and economics (see Table 2.7 for comparisons), its recent energy use at 600.30 kg of oil equivalent per capita and carbon dioxide emissions at 1.67 Mt $CO_2$e/capita both stand abysmally when compared to MDR and even to many of the countries in LDR. In spite of being a developing country, in terms of other indices, such as urbanization, energy and carbon throughput, India occupies the very first quadrant in the 3×3 spatial matrix, making it comparable to some of the LDCs. It is evident that further urbanization and bridging of the urban–rural divide in India poses both a challenge and an opportunity to regulate excessive energy consumption and carbon footprint.

*Table 2.7* India's level of urbanization, emissions, economy and energy use, 1960–2010

|  | *1960* | *1970* | *1980* | *1990* | *2000* | *2010* |
|---|---|---|---|---|---|---|
| Urban population (percentage of total population) | 17.92 | 19.76 | 23.10 | 25.55 | 27.67 | 30.93 |
| Carbon dioxide emissions (tons per capita) | 0.27 | 0.35 | 0.50 | 0.79 | 1.14 | 1.67 |
| GDP per capita at PPP (2005 constant international dollars) | 83.80 | 114.40 | 271.24 | 375.89 | 457.28 | 1,417.06 |
| Energy use (kg of oil equivalent per capita) | – | 275.56 | 293.51 | 364.54 | 438.66 | 600.30 |

Source: UN DESA and the World Bank database, data for energy use unavailable for 1960 and 1970; it is available since 1971, hence 1970 substituted by 1971; all figures rounded off to two decimal points

The analysis demonstrates that India, like many LDR and even LDC societies on development pathways – urbanizing in the range of 34–67% – can continue to develop economically, keeping emissions around 3 t/capita, by controlling their energy consumption levels between 1,000 and 2,000 kg of oil equivalent per capita (as shown by solid arrow (b) against the business-as-usual development trajectory they would follow, and shown with a dotted arrow (a) on the right side in Figure 2.7). This necessitates the urgency to target sustainable urbanization, spatial planning and governance through a low-carbon approach, settlement strategies and mitigation-inclusive instruments, as addressed in subsequent chapters of this book.

## Basic questions

1 Discuss the conceptual relationship between economic development and GHGs. Do income inequalities influence emission inequalities?
2 Differentiate between inequality and inequity. How is the Gini coefficient used to measure income inequalities of nations?
3 Taking the case of developing countries, deliberate the complexities and challenges at the crossroads of urbanization and GHG trajectories.
4 How does the GDP–GHG relationship of nation states compare with their urbanization–GHG relationship?
5 On the basis of energy access/consumption and emissions of countries, comment upon evident global and local inequities.

## Advanced questions

1 Is the North–South divide in climate governance for real? Substantiate your argument on the basis of respective facts and contestations of parties.
2 How can conventional theories and tools, such as EKC, the Lorenz curve and the Gini index, be applied to explain global GHG emissions?
3 Explain the spatial framework to analyse carbon footprints of nations with different economic and urbanization levels.

4    Write a critique of your country's current economic, urban and emission situation. Illustrate their superimposed scenarios for the last four to five decades.
5    Write a commentary on the emerging emission patterns of countries as they grew economically and urbanized in the last half-century.
6    Contemplate how sustainable urbanization could leverage or control national GHG emissions.

## Do it yourself exercises

1    Using the global metadata in Annexure III, prepare your country's GDP–GHG relationship vs. its urbanization–GHG relationship for the last 50 years. Position or compare it with the pattern of 209 countries from 1960 to 2010.
2    Based on Figure 2.10, take your country's case and demonstrate a time-series analysis of urbanization, economic development and GHG emissions.
3    Based on your country's position on the 3×3 spatial development matrix (Figure 2.7), assess what is most optimal – urbanization or development trajectory with the least carbon footprint/impact.
4    Compare the urbanization, economic development and GHG emission status of your country with respect to the UN country classification/grouping or World Bank classification it falls in (alternatively, use Annexure III as reference). Discuss the deviations.

## Suggested reading

### *Global inequities in access and allocation of carbon*

1    Duro, J.A. and Padilla, E. (2006). International inequalities in per-capita $CO_2$ emissions: A decomposition methodology by Kaya factors. *Energy Economics, 28*, 170–87.
2    Stern, N. (2006). *The Stern review on the economics of climate change* (Executive Summary). London: Government Economic Service.
3    Atapattu, S. (2008). Climate change, equity and differentiated responsibilities: Does the present climate regime favor developing countries? Conference on *Climate law in developing countries post-2012: North and South perspectives*. IUCN Law Academy, University of Ottawa, 26–28 September 2008.
4    Purkayastha, P. (2010). Equity and carbon space. Conference on *Global carbon budgets and equity in climate change*. Mumbai, 28–29 June 2010. Retrieved from www.qub. ac.uk/research-centres/RespondingtoClimateChangeIndia-UKPerspectives/Filestore/ Theme1/Filetoupload,239568,en.pdf on 14 March 2014.

### *North–South debates in environmental/climate governance*

1    Najam, A. (2005). Why environmental politics looks different from south. In P. Dauvergne (Ed.), *Handbook of global environmental politics*. Northampton, MA: Edward Elgar.
2    Prum, V. (2007). Climate change and North–South divide: Between and within. *Forum of International Development Studies, 34*.

3  Roberts, J.T. and Parks, B.C. (2007). *A climate of injustice global inequality, North–South politics, and climate policy*. Cambridge, MA/London: MIT Press.
4  Kartha, S., Athanasiou, T. and Baer, P. (2012). The North–South divide, equity and development: The need for trustbuilding for emergency mobilization. *Development Dialogue, 61*, What Next Volume III | Climate, Development and Equity.
5  Pradhan, R.M. (2013). *A critique of North South debate on climate change: The question of the real South*. Retrieved from http://aisc-india.in/AISC2013_web/papers/papers_final/paper_122.pdf on 3 April 2014.

### Spatial framework to analyse carbon footprints across global and local (urban–rural) gradient

1  Sethi, M. and Mohapatra, S. (2013). Governance framework to mitigate climate change: Challenges in urbanising India. In Huong Ha and Tek Nath Dhakal (Eds.), *Governance approaches to mitigation of and adaptation to climate change in Asia* (pp. 200–30). Hampshire, UK: Palgrave Macmillan.
2  Sethi, M. (2015). Location of greenhouse gases (GHG) emissions from thermal power plants in India along the urban–rural continuum. *Journal of Cleaner Production, 103*, 586–600. doi:10.1016/j.jclepro.2014.10.067.
3  Sethi, M. (2015). Mapping global to local carbon inequities in climate governance: A conceptual framework. *International Journal of Development Research, 5*(8), 5387–97.
4  Sethi, M. and Puppim de Oliveira, J.A. (2015). 'North–South' to local 'urban–rural': A shifting paradigm in climate governance? *Urban Climate, 14*(4), 529–43.

## References

ADB (2012). *Key indicators for Asia and the Pacific 2012*. Mandaluyong City, Philippines: Asian Development Bank.
Adger, W.N., Brown, K. and Hulme, M. (2005). Redefining global environmental change (Editorial). *Global Environmental Change: Human and Policy Dimensions, 15*, 1–4.
Allen, M.R., Frame, D.J., Huntingford, C., Jones, C.D., Lowe, J.A., Meinshausen, M. and Meinshausen, N. (2009). Warming caused by cumulative carbon emissions towards the trillionth tonne. *Nature, 458*, 1163–6.
Atapattu, S. (2008). Climate change, equity and differentiated responsibilities: Does the present climate regime favor developing countries? Conference on *Climate law in developing countries post-2012: North and South perspectives*. IUCN Law Academy, University of Ottawa, 26–28 September 2008.
Bulkeley, H. and Newell, P. (2009). *Governing climate change*. New York: Routledge.
Cantore, N. and Padilla, E. (2009). Emissions distribution in post-Kyoto international negotiations: A policy perspective. In N. Ekekwe (Ed.), *Nanotechnology and microelectronics: Global diffusion, economics and policy* (pp. 1–15). Hershey, PA: IGI Global.
Census of India (2011). *Provisional Population Totals 2011*. Paper II, 2. New Delhi: Census of India.
Chakravarty, S., Chikkatur, A., de Coninck, H., Pacala, S., Socolow, R. and Tavoni, M. (2009). Sharing global $CO_2$ emission reductions among one billion high emitters. *PNAS, 106*(29), 11884–8. doi:10.1073/pnas.0905232106.
Climate Justice (n.d.). *Climate Justice/Debt*. Retrieved from http://climate-justice.info/ on 15 October 2015.

Costa, L., Rybski, D. and Kropp, J.P. (2011). A human development framework for $CO_2$ reductions. *PLoS ONE, 6*(12), e29262. Retrieved from http://dx.doi.org/10.1371/journal.pone.0029262 on 25 October 2015.

den Elzen, M.G.J. and Hohne, N. (2008). Reductions of greenhouse gas emissions in Annex I and non-Annex I countries for meeting concentration stabilisation targets. *Climatic Change, 91*, 249–74.

Dhakal, S. (2009). Urban energy use and carbon emissions from cities in China and policy implications. *Energy Policy, 37*(11), 4208–19. doi:10.1016/j.enpol.2009.05.020.

Dhakal, S. (2010). GHG emissions from urbanization and opportunities for urban carbon mitigation. *Current Opinion in Environmental Sustainability, 2*, 277–83.

Duro, J.A. and Padilla, E. (2006). International inequalities in per-capita $CO_2$ emissions: A decomposition methodology by Kaya factors. *Energy Economics, 28*, 170–87.

Eckl, J. and Weber, R. (2007). North: South? Pitfalls of dividing the world by words. *Third World Quarterly, 28*(1), 3–23.

GEA (2012). *Global energy assessment: Towards a sustainable future.* Cambridge/New York: Cambridge University Press & IIASA, Austria.

Groot, L. (2010). Carbon Lorenz curves. *Resource and Energy Economics, 32*, 45–64.

Gupta, J., International Institute for Sustainable Development and Center for Sustainable Development in the Americas (2000). *On behalf of my delegation: A survival guide for developing country climate negotiators.* Winnipeg: International Institute for Sustainable Development.

Heil, M.T. and Wodon, Q.T. (1997). Inequality in $CO_2$ emissions between poor and rich countries. *Journal of Environment and Development, 6*, 426–52.

Heil, M.T. and Wodon, Q.T. (2000). Future inequality in $CO_2$ emissions and the impact of abatement proposals. *Environmental and Resource Economics, 17*, 163–81.

HPEC (2011). *Report on Indian urban infrastructure and services.* New Delhi: The High Powered Expert Committee on Urban Infrastructure constituted by Government of India.

IEA (2011). *$CO_2$ emissions from fuel combustion.* Paris: IEA/OECD.

IEA (2012). *Understanding energy challenges in India: Policies, players and issues.* Paris: IEA.

IPCC (2007). *Climate change 2007: The physical science basis. Contribution of working group I to the fourth assessment report of the Intergovernmental Panel on Climate Change* [S. Solomon, D. Qin, M. Manning, Z. Chen, M. Marquis, K.B. Averyt, M. Tignor and H.L. Miller (Eds.)]. Cambridge/New York: Cambridge University Press.

Jorgenson, A., Clark, B. (2010). Assessing the temporal stability of the population/environment relationship in comparative perspective: A cross-national panel study of carbon dioxide emissions, 1960–2005. *Population and Environment, 32*, 27–41.

Kartha, S., Athanasiou, T. and Baer, P. (2012). The North–South divide, equity and development: The need for trustbuilding for emergency mobilization. *Development Dialogue, 61*, What Next Volume III | Climate, Development and Equity.

Kato, H. (2001). *Regional/sub-regional environmental cooperation in Asia.* Kanagawa, Japan: The Institute of Global Environmental Strategies.

Mahabub-ul-Haq (1976). *The poverty curtain.* New York: Columbia University Press.

Martine, G. (2009). Population dynamics and policies in the context of global climate change. In J.M. Guzman, G. Martine, G. McGranaghan, D. Schensul and C. Tacoli (Eds.), *Population dynamics and climate change* (pp. 9–30). New York/London: UNFPA & IIED.

Mckinsey (2010). *India's urban awakening: Building inclusive cities sustaining economic growth.* New Delhi: Mckinsey Global Institute.

Meinshausen, M. *et al.* (2009). Greenhouse-gas emission targets for limiting global warming to 2° C. *Nature*, *458*, 1158–63.

Miguez, J. and Domingos, G. (2002). Equity, responsibility and climate change. In P.R. Luiz and M. Mohan (Eds.), *Ethics, equity and international negotiations on climate change* (pp. 7–35). Cheltenham, UK: Edward Elgar.

Milanovic, B. (2005). W*orlds apart: Measuring international and global inequality*. Princeton, NJ: Princeton University Press.

MoEF (2006). *National environment policy*. New Delhi: Ministry of Environment and Forests, Government of India.

MoEF (2008). *National action plan for climate change*. New Delhi: Ministry of Environment and Forests, Government of India.

MoEF (2010). *India's GHG emissions profile*. New Delhi: Climate Modelling Forum, Ministry of Environment ans Forests, Government of India.

MoEF (2012). *India: Greenhouse gas emissions 2007*. New Delhi: Indian Network for Climate Change Assessment, Ministry of Environment and Forests, Government of India.

MoUD (2011). *India's urban demographic transition: The 2011 census results – provisional*. New Delhi: JNNURM Directorate and National Institute of Urban Affairs.

Najam, A. (2005). Why environmental politics looks different from south. In P. Dauvergne (Ed.), *Handbook of global environmental politics*. Northampton, MA: Edward Elgar.

Newell, P. (2005). Race, class and the global politics of environmental inequality. *Global Environmental Politics*, *5*(3), 70–94.

Padilla, E. and Serrano, A. (2006). Inequality in $CO_2$ emissions across countries and its relationship with income inequality: A distributive approach. *Energy Policy*, *34*(14), 1762–72.

Parshall, L., Gurney, K., Hammer, S.A., Mendoza, D., Zhou, Y. and Geethakumar, S. (2009). Modeling energy consumption and $CO_2$ emissions at the urban scale: Methodological challenges and insights from the United States. *Energy Policy*, *38*(9), 4765–82. doi:10.1016/ j.enpol.2009.07.006.

Peters, G.P. (2010). Carbon footprints and embodied carbon at multiple scales. *Current Opinion in Environmental Sustainability*, *2*, 245–50.

Potts, D. (2009). The slowing of Sub-Saharan Africa's urbanization: Evidence and implications for urban livelihoods. *Environment and Urbanization*, *21*, 253–9. doi:10.1177/0956247809103026.

Poumanyvong, P. and Kaneko, S. (2010). Does urbanization lead to less energy use and lower $CO_2$ emissions? A cross-country analysis. *Ecological Economics*, *70*, 434–44.

Pradhan, R.M. (2013). *A critique of North South debate on climate change: The question of the real South*. Retrieved from http://aisc-india.in/AISC2013_web/papers/papers_final/paper_122.pdf on 3 April 2014.

Purkayastha, P. (2010). Equity and carbon space. Conference on *Global carbon budgets and equity in climate change*. Mumbai, 28–29 June 2010. Retrieved from www.qub.ac.uk/research-centres/RespondingtoClimateChangeIndia-UKPerspectives/Filestore/Theme1/Filetoupload,239568,en.pdf on 14 March 2014.

Richards, M. (2003). *Poverty reduction, equity and climate change: Global governance synergies or contradictions?* Retrieved from www.odi.org.uk/IEDG/publications/climate_change_web.pdf on 12 March 2014.

Roberts, J.T. and Parks, B.C. (2007). *A climate of injustice global inequality, North–South politics, and climate policy*. Cambridge, MA/London: MIT Press.

Satterthwaite, D. (2009). The implications of population growth and urbanisation for climate change. In J.M. Guzman, G. Martine, G. McGranaghan, D. Schensul and C. Tacoli (Eds.), *Population dynamics and climate change* (pp. 45–63). New York/London: UNFPA & IIED.

Sethi, M. (2015). Mapping global to local carbon inequities in climate governance: A conceptual framework. *International Journal of Development Research, 5*(8), 5387–97.

Sethi, M. and Puppim de Oliveira, J.A. (2015). 'North–South' to local 'urban–rural': A shifting paradigm in climate governance? *Urban Climate, 14*(4), 529–43.

Shah, A. (2013). *Poverty facts and stats*. Retrieved from www.globalissues.org/article/26/poverty-facts-and-stats on 20 April 2014.

South Commission (1990). *The challenges to the South: The report of the South Commission*. Oxford: Oxford University Press.

Stern, N. (2006). *The Stern review on the economics of climate change* (Executive Summary). London: Government Economic Service.

TCPO (2012). *Data highlights (urban) based on census of India*. New Delhi: Town & Country Planning Organisation, Government of India.

UN (2010). *World urbanization prospects: The 2009 revision (CD ROM edition)*. New York: United Nations, Department of Economic and Social Affairs, Population Division.

UN DESA (2012). *World urbanization prospects: The 2011 revision*. United Nations Population Division, Department of Economic and Social Affairs. New York: United Nations.

UNDP/WHO (2009). *The energy access situation in developing countries*. New York: United Nations Development Program & World Health Organization.

UNFCCC (2009). *Ideas and proposals on the elements contained in paragraph 1 of the Bali Action Plan*. Retrieved from http://unfccc.int/resource/docs/2009/awglca6/eng/misc04p01.pdf on 5 May 2014.

UNFCCC (2014). *Kyoto Protocol*. Retrieved from http://unfccc.int/kyoto_protocol/items/2830.php on 2 April 2014.

UN-Habitat, ICLEI and UCLGA (2014). *The state of African cities 2014*. Nairobi: UN-Habitat.

WEO (2008). *World Energy Outlook 2008*. Paris: International Energy Agency.

Wheeler, D. (2011). *Fair shares: Crediting poor countries for carbon mitigation*. Washington, DC: Center for Global Development.

World Bank (2010). *Cities and climate change: An urgent agenda*. Washington, DC: IRDC.

# 3    Role of cities in contributing to national urban GHGs

## Methods, tools and evidence from India

> GHG protocols, standards, methodologies, emission inventories, accounting and metrics (include) – designing, applying and understanding the limitations of different approaches used for measuring, estimating, reporting and verifying GHG emissions and removals.
>
> (Gillenwater, 2015)

As the previous chapter elucidated, it is now evident that there is a strong relation between spatial development such as urbanization and the GHG emissions at the international and national scales. This chapter is in pursuit of the research question of determining the role, both theoretical and empirical, of urban areas in contributing to national GHG emissions, with India being the case in point. This essentially mandates a comprehensive review of existing footprint methodologies, inventories and tools that are used to quantify GHG emissions. In this regard, there is a growing interest in the carbon footprint of cities and geopolitical regions (Hillman and Ramaswami, 2010; Kennedy *et al.*, 2009; Larsen and Hertwich, 2009; Lenzen and Peters, 2010). Since the mid-1990s and the early 21st century, efforts to develop GHG inventories (at the project, national, regional or international levels) have been elevated from an almost purely (and at times obscure) technical matter to a critical and central component of any system or policy that aims to mitigate the effects of climate change.

Drawing on earlier work in the field of air pollutant inventories, the foundation of scientific quantification of GHGs was laid by the IPCC while developing the 1996 and 2006 Guidelines for National Greenhouse Gas Inventories (Pulles, 2011). Resultantly, GHG protocols, standards, methodologies, emission inventories, accounting and metrics include designing, applying and understanding the limitations of different approaches used for measuring, estimating, reporting and verifying GHG emissions and removals (including issues such as boundaries, additionality, baselines, leakage and permanence), and using different technologies for various accounting frameworks and sectors (e.g. fuel combustion, agriculture, forestry, waste management) (Gillenwater, 2011). Due to so many variables, it is reasoned that methods used to determine the carbon footprint should not be specified in the definition. It is only necessary that the method satisfactorily meet the requirements of the definition (Peters, 2010). But in order to account for the contribution of urban areas, countries or any other entity to climate change, it is necessary to quantify their GHG emissions. This requires particular methodologies to account for the various activities and the volume of these gases that they produce (UN-Habitat, 2011). And in order to make meaningful comparisons over time, or between different places, there is a need for standardized protocols to be developed. According to the UNFCCC's protocol, inventories should meet the following five quality criteria (UNFCCC, 2004):

1    *Transparency*: assumptions and methodologies should be clearly explained.
2    *Consistency*: the same methodology should be used for base and subsequent years.
3    *Comparability*: inventories should be comparable between different places.
4    *Completeness*: inventories should cover all relevant sources of emissions.
5    *Accuracy*: inventories should be neither over nor under true emissions.

National inventories are prepared according to a detailed set of criteria developed by the IPCC (IPCC, 2006), discussed in section 3.3, but there is no international protocol or convention in place to measure GHG emissions from subnational areas. In addition to the standard methodologies promoted by the industry, scientific community and international or multilateral organizations, this chapter deliberates into the background of the scientific methods on which these methodologies are based. This is fundamental to the discourse of quantifying GHG emissions, as it distinguishes between actual measurements of emissions versus the estimation approach, which is often misinterpreted as the former in the policymaking community.

The methodologies have distinct relevance to and bearing upon different functional units of interest. Accordingly, they could be classified according to the scale, from global to local, say a product, corporation or a city. Thus, this chapter investigates not only urban relevant methodologies, tools and models, but also the ones at national/country and corporate/business/household levels to infer their influence on developing relevant urban methodologies. This chapter captures their respective characteristics, similarities and observed inconsistencies, to derive an alternative or the most appropriate methodology to estimate the overall GHG responsibility of the urban areas at national/country level that could lead to a much more favourable and timely policy initiative for mitigating actions. The adopted method would eventually employ evidence/data from India to evaluate the quantum of GHGs contributed by its urban areas to the national throughput and report major drivers and activities thereof.

## 3.1 Scientific methods for GHG quantification/carbon footprint

GHG emissions can be quantified either by directly measuring them or by estimating them. Based on a review of the IPCC 2006 Guidelines and guidelines from national government agencies, the World Resources Institute (2002) recognizes four main quantifying methods – the emission factor-based method, the mass balance method, the predictive emissions monitoring system and the continuing emissions monitoring system.

**Emission factor-based method:** This method is often used to estimate the emissions of large entities, such as countries or cities, but it can also be used for small entities. The 'emission factor' is a coefficient that quantifies the emissions per activity. Site-specific data on the exact quantity of GHG emissions are not needed. Instead, data samples are used that represent the amount of GHG emissions released when a certain activity is carried out under specific operation conditions. The factor-based approach can be written as follows: $E = A \times EF$, where $E$ represents the emissions, $A$ represents the activity data (e.g. fuel consumption or production output) and $EF$ represents the emission factor (expressed as a specific value in t $CO_2$/TJ or kg $CO_2$/t). The precondition for using this method is, of course, that emission factors have been calculated for the activity to be measured. As operation conditions differ across countries/sites, it may be necessary to calculate site-specific or local emission factors to improve the accuracy of the measurement.

**Mass balance method:** The basic idea of this method is to follow the mass flow of an element such as carbon or oxygen through a process. This method can be used if the input/output streams, as well as the chemical reactions of a process, can be well identified (e.g. for stationary combustion technologies). Its general equation reads as follows: input = output + emissions.

**Predictive emissions monitoring system (PEMS):** This method comprises elements of the direct measurement and the calculation-based approach. It requires that, for the unit in question, a correlation test is made to determine the relationship between process parameters and the level of GHG emissions. The determined correlation serves as input for mathematical models that calculate the released emissions for a given process.

**Continuous emissions monitoring system (CEMS):** The CEMS approach is based on direct measurement of emissions. It allows obtaining very accurate and real-time data. Depending on the purpose of the GHG measurement, different methods may be used. For a voluntary programme that aims at gathering data on a wide range of gases and emission sources, the emission factor or mass balance method may be very suitable given the various sources. If emissions are to be measured in a regulatory framework, and thus for mandatory purposes, the methods to be used may be already defined. Some protocols and programmes define 'tiers' to indicate different levels of accuracy. Often, three tiers are given whereby the tier 3 method is the most accurate and the tier 1 method the least accurate (see Box 3.1 for more information on this). The IPCC recommends using tier 2 or 3 methods to calculate the key emission sources for a national inventory (IPCC, 2006).

**Box 3.1    About emission factors and tiers**

Many tools provide emission factors in view of more easily rendering the compilation of the inventory. GHG emissions can thus be calculated by multiplying specific activity data (e.g. total gasoline consumption within the territory) by the corresponding emission factor. The IPCC provides default emission factors. The use of these default emission factors would represent a tier 1 approach (i.e. the least accurate emission estimation). A more accurate tier 2 approach requires that default emission factors are replaced by country-specific emission factors that take account of country-specific data. For instance, a country-specific emission factor for fuel combustion would take account of the average carbon content of the fuel, fuel quality, carbon oxidation factors and the state of technology development. A tier 3 approach would, in addition, take account of operation conditions, the age of the equipment used to burn the fuel, control technology, operating conditions, the fuel type used and combustion technology. Such an approach represents the most accurate emission quantification. However, for many local territories, the use of a tier 3 approach might be too complex. For big plants, data on plant-specific $CO_2$ emissions are increasingly available. It is good practice to use the most disaggregated site- and technology-specific emission factors available (IPCC, 2006). If a local authority has access to country and regional emission factors for key activities, then the regional emission factors should be preferred.

Source: IPCC (2006)

**3.2 Application of methodologies: scale-based classification**

In an overview of various methods to quantify emissions, Peters (2010) suggests that the application of a particular method is subject to the scale at which an assessment is to be performed. Accordingly, the methods used to determine the carbon footprint should not be specified in the definition. It is only necessary that the method satisfactorily meet the requirements of the definition. In practice, the method depends on functional unit via the operating scale, which could range from global, country or city-country level to a business, industry, household down to a specific product. Methods to evaluate carbon footprint vary accordingly from input–output models to process- or life-cycle-based assessments at the micro level, and several hybrids between the two extremes, explained diagrammatically in Figure 3.1. It clearly indicates that there could be different methods used to account for emissions at multiple scales.

Peters (2010) and Finkbeiner (2009) uphold that consumer products would generally use the bottom-up life cycle assessment (LCA) method, while studies at the national level would apply top-down input–output analysis (IOA)

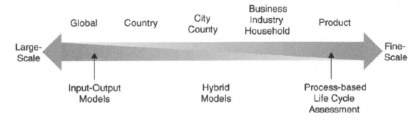

*Figure 3.1* Scheme of carbon footprint applications and corresponding methods across
scales

Source: Peters (2010)

(Wiedmann, 2009). Hybrid methods that combine the strength of both LCA and
IOA are an active area of research and are increasingly being used in practice
(Lenzen, 2009; Suh and Nakamura, 2007; Williams *et al.*, 2009). This invariably takes into consideration the functional requirements and the degree of detail
requisite at that particular functional level. In order to appreciate accounting
methods at the urban scale, it becomes relevant to understand country-/national-
and corporate-/business- or household-level methods. It thus also opens up a
large scope to devise hybrid models for meso-scale entities such as numerous
urban areas within an entire country, which is otherwise unsuitable to assess for
individual cities, owing to inherent conceptual and practical complexities.

### 3.3 Methodologies to estimate country/national emissions

National inventories are prepared according to a detailed set of procedures
developed by the IPCC (IPCC, 2006). The 2006 IPCC Guidelines provide
methodologies for estimating national anthropogenic emissions by sources and
removals by sinks of GHGs. They may assist parties (nations) in fulfilling their
commitments under the UNFCCC on reporting on inventories of anthropogenic
emission of GHGs not controlled by the Montreal Protocol, as agreed by the
parties. The 2006 IPCC Guidelines are in five volumes. Volume 1 describes the
basic steps in inventory development and offers general guidance in GHG emissions and removal estimates based on its authors' understanding of accumulated
experiences of countries over the period since the late 1980s, when national GHG
inventories started to appear in significant numbers. Volumes 2 to 5 offer guidance for estimates in different sectors of economy. These are: energy; industrial
processes and product use; agriculture, forestry and other land use; and waste
(see Table 3.1). It essentially follows the 'emission factor' and 'tier'-based estimation method for each one of the activities or sectors.

The 2006 IPCC Guidelines evolved gradually from the 1996 IPCC Guidelines,
Good Practice Guidance 2000 and Good Practice Guidance LULUCF, and have
gained general acceptance among countries as the basis for inventory development. According to its definition, national inventories of anthropogenic GHG

*Table 3.1* Contents of the five volumes that make up the 2006 IPCC guidelines

| Volumes | Chapters |
|---|---|
| 1 – General Guidance and Reporting | 1  Introduction to the 2006 Guidelines |
| | 2  Approaches to Data Collection |
| | 3  Uncertainties |
| | 4  Methodological Choice and Identification of Key Categories |
| | 5  Time Series Consistency |
| | 6  Quality Assurance/Quality Control and Verification |
| | 7  Precursors and Indirect Emissions |
| | 8  Reporting Guidance and Tables |
| 2 – Energy | 1  Introduction |
| | 2  Stationary Combustion |
| | 3  Mobile Combustion |
| | 4  Fugitive Emissions |
| | 5  $CO_2$ Transport, Injection and Geological Storage |
| | 6  Reference Approach |
| 3 – Industrial Processes and Product Use | 1  Introduction |
| | 2  Mineral Industry Emissions |
| | 3  Chemical Industry Emissions |
| | 4  Metal Industry Emissions |
| | 5  Non-Energy Products from Fuels and Solvent Use |
| | 6  Electronics Industry Emissions |
| | 7  Emissions of Fluorinated Substitutes for Ozone Depleting Substances |
| | 8  Other Product Manufacture and Use |
| 4 – Agriculture, Forestry and Other Land Use | 1  Introduction |
| | 2  Generic Methodologies Applicable to Multiple Land Use Categories |
| | 3  Consistent Representation of Lands |
| | 4  Forest Land |
| | 5  Cropland |
| | 6  Grassland |
| | 7  Wetlands |
| | 8  Settlements |
| | 9  Other Land |
| | 10  Emissions from Livestock and Manure Management |
| | 11  $N_2O$ Emissions from Managed Soils, and $CO_2$ Emissions from Lime and Urea Application |
| | 12  Harvested Wood Products |
| 5 – Waste | 1  Introduction |
| | 2  Waste Generation, Composition and Management Data |
| | 3  Solid Waste Disposal |
| | 4  Biological Treatment of Solid Waste |
| | 5  Incineration and Open Burning of Waste |
| | 6  Wastewater Treatment and Discharge |

Source: IPCC (2006)

emissions and removals consistent with good practice are those that contain neither over- nor underestimates, so far as can be judged, and in which uncertainties are reduced as far as practicable. These requirements are intended to ensure that estimates of emissions by sources and removals by sinks, even if uncertain, are bona fide estimates, in the sense of not containing any biases that could have been identified and eliminated, and that uncertainties have been reduced as far as practicable, given national circumstances. Estimates of this type are presumably the best attainable, given current scientific knowledge and available resources.

The 2006 IPCC Guidelines generally provide advice on estimation methods at three levels of detail, from tier 1 (the default method) to tier 3 (the most detailed method). The advice consists of mathematical specification of the methods, information on emission factors or other parameters to use in generating the estimates, and sources of activity data to estimate the overall level of net emissions (emission by sources minus removals by sinks). Properly implemented, all tiers are intended to provide unbiased estimates, and accuracy and precision should, in general, improve from tier 1 to tier 3. The provision of different tiers enables inventory compilers to use methods consistent with their resources and to focus their efforts on those categories of emissions and removals that contribute most significantly to national emission totals and trends.

With no bias towards the application of the 2006 IPCC Guidelines, Peters (2010) observes that there is a growing number of studies on the carbon footprint of nations, but very few include the regional detail necessary for a correct calculation of the emissions associated with imported goods and services (Wiedmann, 2009). Also, policy-relevant issues, such as carbon leakage, competitiveness concerns, border tax adjustments and the distribution of emissions between countries, are a natural part of carbon footprint analysis (Davis and Caldeira, 2010; Peters and Hertwich, 2008a, 2008b; Weber and Peters, 2009). Of particular interest at the country level is how the carbon footprint changes over time with respect to territorial emissions. More detailed studies across a wider range of countries and time spans will allow a more holistic assessment of the effectiveness of climate policy.

The global results attained for national GHG emissions while applying IPCC methodology hold due relevance for urban areas. In case of urban areas, emissions from the use of fossil fuels, industrial processes and product use and waste are of particular importance. Stationary combustion mainly relates to energy industries, manufacturing industries and construction, while mobile combustion includes transportation emissions from civil aviation, road transportation, railways and waterborne navigation (although only within national boundaries – fuel use associated with international maritime transportation is not included). These distinctions are important, as – taken as a whole – energy, transportation and buildings account for over half of all global emissions (see Figures 1.1a, 1.1b and 1.2 in Chapter 1).

National GHG inventories are based on the assumption that a country is responsible for all emissions produced within its area of jurisdiction. As a pragmatic measure to facilitate national targets and reductions, this is likely to be the only enforceable strategy – as countries only have legislative power (to govern) within their own national boundaries (UN-Habitat, 2011, p. 34). The same principle needs to be honoured while allocating responsibilities for subnational entities such as urban emissions, while reporting and mitigating them. The IPCC Guidelines are intended to help prepare national inventories of emissions by sources and removals by sinks. Nonetheless, it has been suggested within the guidelines that they would also be relevant for estimating actual emissions or removals at the entity or project level (IPCC, 2006, p. 12). It implies that the national inventory will be relevant to city-regions, sub-national regions like cities or states and by entity, it includes any company or corporation like the municipal corporations of a city.

In a nutshell, the implications of international methodology such as the IPCC Guidelines upon urban areas, and city-level methodologies in particular, are threefold:

(a)  Global or national sectoral differentiation of GHG suggests urban contributions, particularly in energy, transport and buildings sector.
(b)  National responsibilities based on the principles of 'territoriality' and/ or "production/generation source of emissions" enshrined in the 2006 IPCC Guidelines forms a firm basis for carbon footprinting methods at the sub-national level.
(c)  It gives a 'systems approach' of assessing emissions from multiple scales (i.e. macro, meso to micro levels) such that there is no leakage and each carbon-emitting activity or parameter in a city is accounted for.

### 3.4 Methodologies to estimate corporate/business and household emissions

Subnational entities include corporate/business organizations and households. It has been observed that studies pertaining to carbon footprint of companies and households gained interest after the energy crisis of the 1970s. Studies have focused on various aspects such as direct energy use at the household level and the footprint owing to goods and services purchased by them. Most of these assessments are based on surveys of expenditure patterns and reveal a strong association with socio-economic characteristics such as household size, spatial location, lifestyles, eating habits, etc. (Baiocchi *et al.*, 2010; Druckman and Jackson, 2009; Hertwich, 2005; Tukker *et al.*, 2010; Weber and Matthews, 2008a).

A recent study in China (Lin *et al.*, 2013) states that up to 70% of household GHG emissions are from regional and national activities that support household consumption, including the supply of energy and building materials. It also reveals that there is a large disparity in GHG emissions profiles among different households, with high GHG emissions households emitting about five times more than low GHG emissions households. This clearly suggests the importance of studying emissions from the domestic or residential sector as a separate source other than industry or high-commercial-based emissions. Dhakal (2010) has also corroborated, while studying energy and emission profiles of Asian cities, that a city-level comparison may mean analysing emissions from non-industrial or commercial sectors.

**Box 3.2    The International Organization for Standardization (ISO) 14064 and its relation to local government GHG management**

ISO 14064-1:2006, another methodology to account carbon footprint of a corporation or product, specifies principles and requirements at the organization level for quantification and reporting of GHG emissions and removals. It includes requirements for the design, development, management, reporting and verification of an organization's GHG inventory. ISO 14064 requirements have implications on local government GHG management programmes (many are used by the WRI/WBCSD GHG Protocol). In order to comply, at least three additional elements should be incorporated by cities:

**Reporting by individual GHG**

The ISO requirement is to report emissions of each of the six GHGs listed in the UNFCCC and Kyoto Protocol. Local governments have traditionally reported total emissions of 'tons $CO_2$e', but should also report by individual GHG in order to satisfy the requirements of ISO 14064. Where a local government does not intend to seek compliance with ISO 14064, it may report total emissions using the unit 'tons $CO_2$e'. Consistent with reporting under the Kyoto Protocol, local governments should follow international convention in using the GWP outlined in the IPCC's Second Assessment Report.

**Uncertainty assessment**

Analysis is subject to a degree of uncertainty that may arise from the accuracy of available activity data, the degree of alignment with annual or monthly time periods, or variation in the quality of fuel sources. To facilitate accreditation to the ISO 14064 standards, a local government should assess the uncertainty associated with an inventory/analysis of any given year. Local governments wishing to include an uncertainty estimate should refer to the 2006 IPCC Guidance, Chapter 3, "Uncertainties for detailed methods".

**Document retention**

An initial analysis is usually conducted for a historical year, in order to generate baseline data. A subsequent analysis may be done up to five years later than the initial exercise. To enable auditing, all original documents must be retained by the local government.

In addition to regulatory compliance, industries and corporations are gradually becoming aware of the impacts that their activities have upon the local and global environment, by which they have increasingly engaged in conducting GHG inventories. One of the oldest and most commonly used inventory by companies is the WRI/WBCSD GHG Protocol. This protocol puts forward an accounting system that is based on relevance, completeness, consistency, transparency and accuracy, and provides a mechanism by which private sector actors can contribute to the global goal of reducing GHG emissions (WRI/WBCSD, 2013). The inventory also describes the processes by which GHG emissions can be identified and calculated, how these can be verified for a formal reporting process, and how targets can be set. In this process, the concept of 'Scope' was developed further, which takes into account direct and indirect GHG emissions in a more effective manner.

Scope 1 are the emissions from sources under the jurisdiction, owned or controlled of the company. (e.g. emissions from combustion in owned or controlled boilers, furnances, vehicles, or emissions from chemical production in process equipment). Scope 2 are offsite emissions from the purchase of electricity, consumed by the company. Scope 3 are offsite emissions from the company's activities, such as supply chain or from products sold by the company, but sources not owned or controlled by the company. In the GHG Protocol, Scope 3 emissions are not mandatory, but a robust carbon footprint requires all three components.

While emissions from Scopes 1 and 2 are largely estimated from emission factor and activity-based method, emissions from Scope 3 largely rely upon LCA-based method or analysis of products and supply chain. It is believed that Scope 3 emissions are the most difficult for a company to estimate, but, as for household consumption, they are often the most important (Huang *et al.*, 2009; Matthews *et al.*, 2008). Hence, the failure to include Scope 3 emissions can lead to perverse incentives such as outsourcing activities to different companies (shifting emissions from Scope 1 to Scope 3). While Scope 3 emissions are perceived as overly difficult for an average company to calculate, the tools and methods are certainly available (Lenzen *et al.*, 2007; Wiedmann *et al.*, 2009). As for LCA, various hybrid methods can be used to improve the initial estimate practice (Lenzen, 2009; Suh and Nakamura, 2007; Williams *et al.*, 2009).

Methodologies to estimate corporate/business and household hold immense importance and potential to assess carbon footprint of subnational entities at urban-scale or the city. Municipal corporations, development authorities, public, private, PPP organisations, special purpose vehicles, etc. or companies are established with a specific mandate to develop or execute a special project in land development, infrastructure, energy, transport, waste management, civic services, etc. As such, methodologies find practical application in measuring impacts, offsets, trade-offs, co-benefits during city planning, project planning, implementation, monitoring, etc.

## 3.5 Methodologies to estimate city/urban emissions

City authorities often function at two distinct levels. First, they function as business enterprises – owning or leasing buildings, operating vehicles, purchasing goods and carrying out various other activities. In this regard, urban authorities can assess their emissions as corporate entities, including the direct and indirect impacts of their work. Second, urban authorities also function as governments – with varying levels of oversight for and influence on the activities taking place within the spatial area over which they have jurisdiction. There are certain limitations while directly utilizing either the IPCC-based country-level inventories or corporate protocols for urban assessments. For instance, the IPCC methodology for countries does not provide specifications at the local authority level for discussing energy consumption, transportation or waste disposal, and the existing corporate accounting protocols do not cover the details of municipal operations such as street lighting, landfill emissions and emissions from waste water treatment or other industrial activities (UN-Habitat, 2011). Accordingly, the methods to estimate GHG emissions from local governments have gradually evolved from both IPCC methodology for national inventories and protocols for estimating corporate/business GHGs.

City-based studies have used a variety of methods and data to account GHGs (Dhakal, 2010). These include a combination of sales data (e.g. for oil, gas and electricity), estimated levels of activities (trip surveys and household surveys to generate average activity levels), scaling from regional and national information, and modelling. Their treatment to aviation, marine and road transport, those sectors that interact beyond city boundaries, vary from city to city, adding to the existing complexities (Kennedy *et al.*, 2009). However, in contrast to the IPCC's territorial accounting principles, several research methods and climate action plans of cities allocate the electricity-related $CO_2$ emissions to the city's carbon estimates. The existing studies differ methodologically, and there are inconsistencies in respect of gases measured, emission sources covered, sector definitions, scopes of measurement, global warming potential, and IPCC tier methods (Bader and Bleischwitz, 2009; Kennedy *et al.*, 2009). Before adopting any particular methodology in this research, it would be pertaining to understand the plethora of methods, protocols and tools, etc. available, their relevance and shortcomings in application.

Based on the origin, city-/urban-based carbon or GHG emission methodologies could be classified into three main categories: (1) international standards and protocols; (2) evidence in peer-reviewed methodologies/inventories; and (3) computer models and tools.

## 3.6 International standards and protocols for city-based GHG emissions

Although there is no legally accepted international protocol by nations, in practice, there are three international standards/protocols for estimating city or local urban GHG emissions, which are listed in chronological order of their origins along with a summary of their features in this section.

- International Local Government GHG Emissions Analysis Protocol, 2009
- International Standard for Determining GHG Emissions for Cities, 2010
- Global Protocol for Community-Scale Greenhouse Gas Emissions, 2012

The most common feature in all the standards is that they borrow the concept of Scope (Scope 1, Scope 2 and Scope 3) from the WRI's GHG Protocol for companies (explained in section 3.4). The three classifications used to categorize emissions sources, differing slightly when applied in the context of government operations and community-scale inventories, are:

- *Scope 1 emissions*: Direct emission sources owned or operated by the local government (e.g. a municipal vehicle powered by gasoline or a municipal generator powered by diesel fuel).
- *Scope 2 emissions*: Indirect emission sources limited to electricity, district heating, steam and cooling consumption (e.g. purchased electricity used by the local government, which is associated with the generation of greenhouse gas emissions at a power plant). These emissions must be included in the government operations analysis, as they are the result of the local government's operations and energy purchasing policies.
- *Scope 3 emissions*: All other indirect and embodied emissions over which the local government exerts significant control or influence (e.g. emissions resulting from contracted waste hauling services ought to be accounted for).

### 3.6.1 International Local Government GHG Emissions Analysis Protocol (IEAP)

The IEAP was developed in 2009 by the International Council for Local Environmental Initiatives (ICLEI) to take into account both government and community sectors, using the main categories derived from the IPCC's guidelines on national inventories of GHG emissions from stationary combustion, mobile combustion, fugitive emissions, product use, other land use and waste (ICLEI, 2008). It was developed to provide an easily implementable set of guidelines to assist local governments in quantifying the GHG emissions from both their internal operations and from whole communities within their geopolitical boundaries.

ICLEI utilized its experience of working with hundreds of local governments on climate action planning in the framework of Cities for Climate Protection (CCP). Initiated in 1993, the CCP was perhaps the first international initiative that aimed to facilitate emissions reduction of local governments through a five milestone process: (1) preparation of baseline inventory; (2) commitment; (3) planning; (4) implementing; and (5) monitoring using tailored tools such as the Harmonized Emissions Analysis Tool (ICLEI, 2014).

These principles have been adapted from the WRI/WBCSD GHG Protocol Initiative to apply accounting and reporting of GHG emissions, and at the same time follow the IPCC principles of TCCCA. The inventory defines GHGs and biological versus fossil carbon sources, and adheres to the GWP of the IPCC's Second Assessment Report. While defining the boundary of an organization at the local level, it considers using an extension of the concept of control and influence, which recognizes the broad role of local governments as entities that provide services and develop policy affecting the local community. The sectors used in a local government's GHG management reflect their operations and the way in which they interact with their communities. Simultaneously, this local government GHG analysis protocol conforms to the international standards for national and corporate reporting to ensure consistency and comparability.

### 3.6.2 International Standard for Determining GHG Emissions for Cities

The International Standard for Determining Greenhouse Gas Emissions for Cities (UNEP *et al.*, 2010) provides a common method for cities to calculate the amount of GHG emissions produced within their boundaries. This standard builds on, and is consistent with the IPCC protocols for national governments, and provides a common format to facilitate compilation by local authorities. This standard is one of the several tools for cities and climate change developed jointly by UNEP, UN-Habitat and the World Bank, and is supported by Cities Alliance. City baselines developed using methodology that is consistent with this standard are now available for approximately 50 cities.

While measurement should not delay action, a critical requirement to support policy and access to finances is the establishment of an open, global and harmonized protocol for quantifying the GHG emissions attributable to cities and local regions. The purpose of this agreement is to establish a common standard by which tools to inventorize the city emissions should be based upon. It mandates that with the exception of territorial attribution, GHG inventories for cities should use the principles and methods developed by the IPCC. For instance:

- Inventories should adhere to TCCCA principles. They should be sufficiently disaggregated and consistent to enable effective policy development.
- The most recent IPCC guidelines define sectors contributing to emissions as: energy (stationary and mobile sources); industrial processes and product use (IPPU); agriculture; AFOLU, where significant; and waste.
- Annual calendar year emissions for all six Kyoto gases (i.e. carbon dioxide ($CO_2$), methane ($CH_4$), nitrous oxide ($N_2O$), hydrofluorocarbons (HFCs), perfluorocarbons (PFCs) and sulfur hexafluoride ($SF_6$)), and other GHGs as relevant, in terms of carbon dioxide equivalents.

This standard also recognizes that cities give rise to the production of GHG emissions outside their urban boundaries. This standard duly follows the WRI/WBCSD's *GHG Protocol: A Corporate Accounting and Reporting Standard*, by including out-of-boundary emissions that are driven by activities in cities (see Scopes). Thus, the scope of emissions included in the city GHG Standard produced by UNEP, UN-Habitat and the World Bank includes all emissions produced within a city, major emissions from consumption within a city, and major upstream emissions that are attributable to city residents (World Bank, 2010).

According to this standard, GHG emissions embodied in the food, water, fuels and building materials consumed in cities should also be reported as additional information items. This list of embodied emissions in key urban materials follows from the work of Ramaswami *et al.* (2008). It is argued that the methodology developed by Bilan Carbone (www.ademe.fr/bilan-carbone) could also be used to quantify embodied emissions. This is to avoid policies or actions that lower emissions inside of cities, but at the expense of greater emissions outside. This method also includes a standard reporting format for GHG emissions from cities, which includes information on emission factors and activity levels used in the calculation of emissions. All cities or urban regions with populations over 1 million persons are encouraged to use this reporting standard. It opines that cities with populations below 1 million may use less detailed reporting tables, such as those developed by the European Commission for the Covenant of Mayors.

To be pragmatic, cities may follow the IPCC guidelines (2006 IPCC Guidelines for National Greenhouse Gas Inventories, Volume 1, Chapter 4) for identifying and reporting on key categories of emissions, the sum of which represent at least 95% of total emissions. In many cases, AFOLU and IPPU emissions for cities may be too insignificant to report. The determination of urban GHG emissions by this standard does not imply that local governments are responsible for these emissions. Rather, the inventory reflects the carbon dependence of the urban economy and highlights the extensive experience that local governments already have in monitoring GHG emissions.

### 3.6.3 Global Protocol for Community-Scale GHG Emissions, 2012

The Global Protocol for Community-Scale Greenhouse Gas Emissions (GPC) provides requirements and guidance for cities on preparing and publicly reporting a GHG emission inventory. The primary goal is to provide a standardized step-by-step approach to help cities quantify their emissions in order to manage and reduce their GHG impacts. Development of the GPC commenced in June 2011, as a result of a Memorandum of Understanding between C40 and ICLEI. Thereafter, the two worked in close consultation with local governments, WRI and the joint work programme of the Cities Alliance between the World Bank, UNEP, and UN-Habitat for the standard (discussed above in section 3.6.2).

The GPC is intended for adoption by local authorities or city governments who exercise jurisdiction over a defined geographic area. The GPC can also be useful for subnational entities such as towns, districts, counties, prefectures, provinces and states pursuant to appropriate modifications. The GPC attempts to build upon the knowledge, experiences and practices defined in previously published protocols and standards. These include the International Local Government GHG Emissions Analysis Protocol and the International Standard for Determining GHG Emissions for Cities, among others. The GPC claims that upon full execution, its pilot version would replace the provisions related to the above two protocols. The GPC aims for a low-carbon city planning cycle with six steps: (1) base year inventory; (2) future scenario analysis; (3) target setting; (4) action plan; (5) implementation; and (6) tracking performance. It recommends the following three versions:

- *GPC 2012 Basic*: Covers all Scope 1 and Scope 2 emissions of stationary units, mobile units, wastes and IPPU, as well as Scope 3 emissions of the waste sector. In reporting the total by Basic, Scope 1 emissions from energy generation (GPC I.3.1) are not included in order to prevent double-counting, since the total Basic figure also includes Scope 2 emissions. However in reporting by 'Scopes', total Scope 1 emissions must also include Scope 1 emissions from energy generation.
- *GPC 2012 Basic+*: Covers GPC 2012 Basic, as well as AFOLU and Scope 3 emissions for mobile units.
- *GPC 2012 Expanded*: Covers the entirety of Scopes 1, 2 and 3 emissions, including transboundary emissions due to the exchange/use/consumption of goods and services.

The GPC Pilot Version 1.0 for 30+ cities across the globe is an important advancement in the evolution of community-scale GHG accounting and reporting. This iteration is based primarily upon a production-based inventory. The GPC Full Version 1.0 is propounded to establish a single minimum global standard for cities of all sizes and geographies, although a few limitations in the Pilot Version include: the lack of international consensus on methodologies for accounting the emissions from cross-boundary transportation and AFOLU; possible double-counting between in-boundary power plant emissions and emissions from grid electricity (between Scope 1 and Scope 2); and the lack of framework and methodologies for quantifying full Scope 3 emissions.

### 3.6.4 Limitations of international standards/protocols for city-based GHGs

The most significant limitations observed in all the three standards/protocols is their acute dependence on immaculate scoping and data, which is neither able to give fast results for cumulative urban contributions to GHG within a country, nor able to guide climate policy, low-carbon solutions, inter-sectoral co-benefits, or the national urbanization policy. For instance, IEAP itself proclaims that it is often impractical to define these kinds of attributed emissions sources consistently in different countries, so most Scope 3 emissions are not required in the Global Reporting Standard. It further states that they are discouraged from including Scope 3 sources that are academic and have no obvious policy relevance so the local government does not waste its time collecting irrelevant information.

Similarly, the International Standard for Determining GHG Emissions for Cities though considers different scopes to estimate out of boundary and gateway related emissions. Nonetheless, it recognizes that it is impractical to quantify all of the emissions associated with myriad of goods and materials consumed in cities. On similar lines, the GPC also notes its limitations include the lack of international consensus on methodologies for accounting the emissions from cross-boundary transportation and AFOLU, possible double-counting in Scope 2 and Scope 3 emissions (between in-boundary power plant emissions and emissions from grid electricity), and the lack of framework and methodologies for quantifying full Scope 3 emissions.

Under the above circumstances, it is highly doubtful to arrive at a reasonable or credible figure of urban emissions within a country without accounting. In addition, all the above protocols are at the mercy of voluntary adoption by individual cities. It needs to be underscored that most of the emissions emanating from Scope 3 sources may not have any policy relevance, owing to legal issues or non-jurisdiction of the local government over these sectors or geopolitical areas, such as emissions from thermal power plants or the import of goods being produced outside the city.

## 3.7 Evidence in peer-reviewed methodologies/inventories

In the recent past, there has been growing peer-reviewed literature on methodologies and inventories to quantify GHG emissions at the urban scale. Bader and Bleischwitz (2009), Chavez and Ramaswami (2011), Dhakal (2004, 2010), Dodman (2009), Kennedy *et al.* (2009) and Satterthwaite (2008) have contributed theoretically and/or empirically to enlighten on various aspects of accounting a city's footprint. Though individual cities were constantly studied for their carbon footprint in the past two decades (Toronto, Barcelona, Tokyo, Beijing, etc.), Satterthwaite (2008) was one of the first investigations that conceptually formulated the contribution of urban areas across the globe to cause GHG emissions, pegging their carbon footprint at 30.5–40.8% of the total, based upon a production perspective.

Dodman (2009) was one of the first studies to compare per capita GHG emissions from various world cities. He argued that cities are often blamed for high levels of GHGs, however an analysis of emission inventories shows that in most cases, per capita emissions from cities are lower than the average for the countries in which they are located. The paper assesses these patterns of emissions by cities and by three main sectors – industrial activity, transport and waste – discusses the implications of different methodological approaches to generate inventories, identifies the main drivers for high levels of greenhouse gas production, and examines the role and potential of cities to reduce global GHG emissions.

Kennedy *et al.* (2009) was a first-of-its-kind research that collated and compared GHG emissions for a sample of 44 global cities and metropolitan regions from diverse sources. It reviews different methodologies used to attribute GHGs to urban areas and observes that all are essentially adaptations or simplifications of the IPCC guidelines, while incorporating the WRI/WBCSD concept of Scope 2 and 3 'cross-boundary' emissions. Analysis of previous studies showed that specific differences exist in methodology, for instance some Scope 3 emissions such as those embodied in materials, food, and fuel consumed in cities have only been quantified in a few studies, and thus need to be included in further studies. Elsewhere, baseline emissions are presented with and without emissions from industrial processes, AFOLU (both of which may be incomplete), as well as waste and aviation/marine (for which there are once again differences in adopted methodology). Despite these minor differences, it acknowledges that a potential clearly exists to establish an open, global protocol for quantifying GHG emissions attributable to urban areas.

Dhakal (2010) searches for answers to two broad questions: (1) What do we know about GHG emissions from urbanization at multiple scales? (2) What are the key opportunities to mitigate them from cities and their efficient governance? The review suggests that the quantification of urban contribution to global, regional and national GHGs are limited to a few regions and for $CO_2$ only. The GHG emissions of urban areas differ widely for the accounting methods, Scope of GHGs, emission sources and urban definition, thus making place-based comparisons difficult. Previously, works of Dhakal (2004) concentrated on urban energy use in Asian megacities, where it was reasoned that the nature of energy use and GHG emissions from cities was not well understood in Asia. Energy management at city level was neither a priority nor an important issue until recently because energy-related decisions were made at the national level. These days, city policymakers are under growing pressure to incorporate GHGs, especially $CO_2$ emissions, into consideration while planning. But any policy measure solely for $CO_2$ reduction is a distant possibility for cities in Asia, with the exception of selected and relatively developed cities. In this context, Dhakal (2004) made an attempt to quantify $CO_2$ emissions from energy use and analyse their driving factors for the selected Asian megacities of Tokyo, Seoul, Beijing and Shanghai. Further, it highlights the need to take into account the overall energy and $CO_2$ footprint of cities. Finally, it presents policy directions and policy challenges to identify major opportunities and barriers in integrating $CO_2$ considerations into local environmental policies. A compilation of some notable studies accounting urban GHG emissions worldwide, with comparative details of emission types and methods considered, is listed in Table 3.2.

*Table 3.2* Important peer-reviewed studies accounting GHG emissions worldwide

| Source | Number of cities | Spatial distribution | Emission types considered | Method |
|---|---|---|---|---|
| Kennedy et al. (2009) | 10 | Global | Energy use (residential), industry (commercial energy use), transportation (without shipping and aviation), AFOLU, waste, aviation and/or marine sources (shipping) | Not specified |
| Dhakal (2008) | 3 | China | Energy use (residential) | Energy use, emission factors of fuel types |
| Brown et al. (2008) | 99 | USA | Transportation (without shipping and aviation), energy use (residential) | National databases (passenger and freight transportation, residential energy consumption) |

*(continued)*

*Table 3.2  (continued)*

| Source | Number of cities | Spatial distribution | Emission types considered | Method |
|---|---|---|---|---|
| Dore *et al.* (2006) | 31 | UK | Transportation (without shipping and aviation), energy use (residential), industry (commercial energy use), AFOLU – agriculture, forestry and other land use, aviation and/ or marine sources (shipping) | NAEI UK (emission factor human activity and reported point source emissions) |
| Carney *et al.* (2009) | 17 | Europe | Energy use (residential), industry (commercial energy use), waste, agriculture, forestry and other land use (AFOLU) | GRIP (Greenhouse Gas Regional Inventory Protocol) Europe |
| ICLEI (2009) | 53 | Asia | Transportation (without shipping and aviation), energy use (residential), industry (commercial energy use), waste | HEAT (Harmonized Emissions Analysis Tool) |
| Sovacool and Brown (2010) | 9 | Global | Transportation (without shipping and aviation), energy use (residential), industry (commercial energy use), AFOLU, forestry and other land use, waste | Calculation based on previously published literature |

Source: Adapted from Postdam 2014 (www.pik-potsdam.de/cigrasp-2/city-module/ghg/index.html, accessed 22 January 2014)

## 3.8 Computer tools and models

There are various models or tools, essentially based on computer software, available on the Internet to account for city-based carbon or GHG emissions. They have been developed in the last decade or so, with a user-friendly Excel- or Windows-based interface to enable local governments, environmental NGOs and researchers to estimate urban emissions across various sectors. A review article, based on comprehensive scientific research (Bader and Bleischwitz, 2009), published in *S.A.P.I.EN.S.* (Institute Veolia Environment), looked into how various methodologies used in urban GHG inventories or tools differ. What were the critical variables explaining differences between inventories and whether different GHG inventory tools were compatible – and/or interoperable – and under which conditions? The six analysed tools were: *$CO_2$ Grobbilanz, Eco2Region, The Greenhouse Gas Regional Inventory Protocol (GRIP), Bilan Carbone, $CO_2$ Calculator* and *Project 2 Degrees.*

The *$CO_2$ Grobbilanz* has been widely used in Austria and was developed by the energy agency of the Austrian regions (Energieagentur der Regionen). *Eco2Region* was developed by Ecospeed and has been widely used in different European countries (e.g. Germany, Switzerland, Italy) by members of the international city network 'Climate Alliance'. *GRIP*, a tool developed by the University of Manchester, has been used in metropolitan regions in several European countries. *Bilan Carbone*, developed by the French energy/environmental agency, Agence de l'Environnement et de la Maîtrise de l'Energie (ADEME), has been used by numerous French cities and local governments. The *$CO_2$ Calculator*, developed by the Danish National Environmental Research Institute (COWI) and Local Government Denmark, has been widely used in Denmark. The tool of *Project 2 Degrees* was developed by the Clinton Climate Foundation, Microsoft and the international city alliance ICLEI, and has been used by some members of the C40 city alliance. The analysis of these tools was based upon methodological guidance documents, test versions of the tools and semi-standardized interviews with their developers. The comparison was done on the principle of TCCCA enshrined within 2006 IPCC Guidelines for national GHG inventories. The research concluded with substantial evidence that the methodologies are not consistent, thus making local GHG inventories and tools hardly comparable.

Bader and Bleischwitz (2009) note that an inventory that takes only direct emissions into account is neither consistent nor comparable with an inventory that takes direct and indirect emissions into account. The issues pertaining to differences in interpretation of boundaries and scopes have also been highlighted. In light of these differences with regard to completeness, consistency and accuracy, the authors concluded that today's inventories compiled with different tools are hardly comparable. *It is opined that the root of the problem lies not just in the trade-off between accuracy of the inventory and availability of human resources mentioned by the authors, but essentially owing to multitude of variables involved for different boundaries, scopes, activity data and emission factors and non-integration of urban spatial attributes that are physical and directly measurable.*

The above reasserts the fact that GHG inventories based on emission factors and activity data are highly time- and resource-intensive, and yet do not form a genuine basis for comparison between cities quantified, while using different inventories. While asserting for greater comparability and the need to have a global protocol with agreement on gases to be measured, global warming potential values, emission sources, sector definitions and scopes of measurement (discussed in detail in section 3.9), the authors suggest that it is very unlikely that there will be a common tool, a common standard or communication between tools if the developers and users are not willing to support this process and are not involved in it. This inevitably suggests that there is plenty of scope for other methods and metrics, such as ones associated rather with a city's physical parameters, to be employed for quantification of GHG emissions.

## 3.9 Major inconsistencies in inventories, tools and methodologies

In order to have an internationally accepted standard or protocol for city-based emissions that is readily usable by urban experts, administrators and spatial planners, it is needful to address several inconsistencies in existing inventories, tools and methodologies. Following a process-based classification, they could be broadly grouped into three main types: input, method and output, as represented in Table 3.3.

### 3.9.1 Definition of 'urban' and 'spatial boundary'

It is a known fact that efforts to develop a standardized globally comparable methodology for GHG emissions at the local or municipal level are made more complicated by the wide range of boundary definitions used for these areas (Dhakal, 2010; Dodman, 2009; UN-Habitat, 2011). In general, the smaller the scale, the greater the challenges posed by 'boundary problems', in which it is increasingly difficult to identify which emissions ought or ought not to be allocated to a particular place (Kates *et al.*, 1998). The importance of boundary definitions is clear from studies of urban populations, where differences in how governments define city boundaries have direct effects on spatial structure. For instance, it has been shown how eight different lists of the world's 20 largest cities vary, with only nine cities appearing on all eight lists, and with four different areas competing for the first two ranks (Forstall *et al.*, 2009).

Dhakal (2010) highlights the importance of paying attention to the definitions of the urban population, the urban extent and the city. Depending on the definition used, urban energy and carbon estimates can vary substantially. The estimate for global urban emissions by the IEA is compatible with the UN definition of urban population, in principle. However, one has to note that the UN itself accepts national definitions that vary widely across countries. Even within the same context, diverse urban definitions could lead to variations in carbon throughput. For example, for the US, 76% of the direct final energy consumption occurs in "census urban areas", 59% occurs in 'urbanized areas' and 17% occurs in 'urban clusters' (Parshall *et al.*, 2009). The population figures for some large

*Table 3.3* Three types of inconsistencies in inventories, tools and methodologies

| Input | 1 | Definition of 'urban' and 'spatial boundary' |
|---|---|---|
|  | 2 | Spatial parameterization |
|  | 3 | Data unavailability |
| **Method** | 4 | Sources and sinks covered/sectors |
|  | 5 | Gases and global warming potential |
|  | 6 | Scope – production vs. consumption approach |
|  | 7 | Tiers/uncertainty in measurement |
| **Output** | 8 | Metric/unit of measurement |
|  | 9 | Purpose – optimization or simulation, policy relevance – regulatory, provision, behavioural changes, etc. |

Source: Author

cities are for people living within long-established city boundaries enclosing areas of only 20 to 200 square kilometres, whereas for others (particularly in China) this includes regions with many thousands of square kilometres and a significant rural population (Satterthwaite, 2007).

These complications related to different definitions of cities and urban areas, and different conceptions of the spatial extent of these, are all equally relevant while identifying their GHG emissions. Likewise, energy consumption in urban areas in the US can vary between 37% and 81%, depending on how these areas are defined and bounded in space. Thus, even within a single country, the potential contribution of urban areas to climate change can vary by a factor of two, depending on the spatial definition of these areas (UN-Habitat, 2011, p. 36). It is also argued that the question about the relevant boundaries of a city has to do with the unit to measure – strict city boundaries or the metropolitan area. A metropolitan or functional limit of the city may be the best scale to use, especially for larger cities (World Bank, 2010, p. 21). Meanwhile, most of the GHG or carbon footprints at the city scale are available for the administrative boundaries, which have more geopolitical or territorial basis, rather than a functional one.

### 3.9.2 Spatial parameterization

Comparisons between methodologies and tools suggest a large variation in understanding various spatial parameters, such as physical extent of the city (discussed above), its built form, economic sectors that determine its land use, etc. While comparing tools, it has been highlighted that the overall GHG emissions of a city with heavy industry cannot be compared with the overall emissions of a city based on services (Bader and Bleischwitz, 2009). Dhakal (2004) also detects a similar inconsistency while comparing emissions from Asian megacities. However, the residential sector or the transport sector may well be comparable between two different cities.

Serious anomalies associated with metrics arise while analysing city emissions on a per capita basis. Bader and Bleischwitz (2009) report that absolute figures for GHG emissions are often not comparable, even if sectors are defined in the same way, since the population of the cities or the industrial production differs. In view of rendering inventories, more comparable ratios are often developed (e.g. GHG emissions per square metre in the buildings sector, or emissions per passenger in the public transport sector). Yet, even if common ratios are used, inventories are not easily comparable across cities. For instance, the GHG emissions of the transport sector are, to a large extent, dependent upon the density of the city since there is a correlation between urban population density and GHG emissions from transport.

### 3.9.3 Data unavailability

The accuracy of an inventory largely depends, to a large part, on the data that feeds into it. This raises the question whether cities utilize comprehensive and reliable data sets for the compilation of a GHG inventory. City-level emission analysis generates a variety of data and logistical problems. For instance, there are large information gaps (particularly in developing countries); different information is available at different geographic levels; and political boundaries of cities may change over time, and often include both rural and urban populations – as is the case for Beijing and Shanghai in China (Dhakal, 2004).

The issue of completeness and accuracy is also elucidated in the case of city inventories. Bader and Bleischwitz (2009) note that, unlike nations, cities face a trade-off between completeness of the inventory (in terms of covering all the IPCC subcategories) and data availability/feasibility. Some subcategories of the IPCC guidelines require very specific information indeed. While for a national inventory the accuracy of the data on emissions from "fishing activities (mobile combustion)" may be of some importance, it can be argued that in the case of a city without access to the sea or major lakes, these emissions are likely to represent only an extremely small part of the overall emissions. Given that the availability of data on this or similar activities is often very limited, and the per-ceived share of their emissions is likely to be negligible, cities may refrain from including all possible landuse related activities in their inventory. Moreover, it would not be difficult to believe that, instead, cities concentrate on the main GHG emitting sectors.

### 3.9.4 Sources and sinks covered/sectors

Sectors are defined as the aggregation of specific emission sources (see sec-tion 3.3 for IPCC defined sectors). There are many inventories that either do not account for certain sectors (sources or sinks), LULUCF being the most significant one, or else use different nomenclatures for them. While comparing tools, Bader and Bleischwitz (2009) report that LULUCF, for instance, is only covered by one of the six inventory tools. It needs to be emphasized that the sink and storage capacity of urban areas may be significant, relative to forests (Churkina *et al.*, 2009), in which case there is a strong argument to report sinks and storage within the system boundary (Peters *et al.*, 2009).

Similarly, there are common discrepancies while defining the transport sector. The emissions of the transport sector could, for example, be defined as aviation emissions + emissions of cars + emissions of trucks + emissions of buses + emissions of railways, etc. Sector-specific emissions can only be compared if the sectors are defined in exactly the same way (i.e. cover the same emission sources). In some cases, only the transport of persons and freight transport on road is taken into account, while in other cases all modes of transport (road, rail, marine and air transport) are covered for both persons and freight. To this there is added a slight methodological difference: the estimation of road transporta-tion emissions can be based either on the average mileage of vehicles or on fuel consumption. Hence, IPCC guidelines could provide a common base for local inventories (Bader and Bleischwitz, 2009).

### 3.9.5 Gases and global warming potential

Greenhouse gases are covered in methodologies and inventories in a varying degree. Some cover only $CO_2$ from energy use, while others cover GHGs other than $CO_2$ (for detailed analyses on GHG emissions, emission sources, emission types and methodological differences of 44 cities, see Kennedy *et al.*, 2009).

Another study on tools analyses that the number and type of greenhouse gases included in the different inventories varies widely (Bader and Bleischwitz, 2009). Three out of the six tools analysed take account of all the six gases of the Kyoto Protocol, while two tools allow the taking into account of carbon dioxide, methane and nitrous oxide. The values for one tool are in brackets because the number of GHGs included in the inventory depends on the version of the tool used. The basic version of the tool in question is the most widely used version and covers only carbon dioxide. The advanced versions, however, cover all the six Kyoto gases.

In view of making the climate impact of different GHGs comparable, they are normally converted to $CO_2$ equivalents. $CO_2$ is thereby the reference gas against which other gases are measured, and has a GWP of 1. The GWP represents how much a certain mass of a gas contributes to global warming compared to the same mass of $CO_2$. It is based on the different times "gases remain in the atmosphere and their relative effectiveness in absorbing outgoing thermal infrared radiation" (IPCC, 2007). For instance, nitrous oxide is 310 times more potent than $CO_2$. A ton of nitrous oxide can thus be converted to $CO_2$ equivalents by multiplying it by 310.

The inventories and tools also differ with regard to the global warming potential values that underpin the calculation of $CO_2$ equivalents. The most widely used global warming potential values are still those of the IPCC's Second Assessment Report of 1995. This is mainly due to the fact that the Kyoto Protocol of 1997, and therefore also the respective national inventories, are based on these values. One tool, *Project 2 Degrees* is preloaded with the values of the three last assessment reports. The developers of this tool, however, recommend using the values of the 1995 report given their widespread use on the national level. Two tools, *$CO_2$ Grobbilanz* and *$CO_2$ Calculator*, use the values of the Third Assessment Report for the calculation of $CO_2$ equivalents, whereas only the *Bilan Carbone* tool is based on the Fourth Assessment Report.

The relatively widespread use of the values of the Second Assessment Report is due to two reasons. First, the inventories using these values are consistent, and, second, to a certain extent also comparable with national inventories. However, with regard to the requirement of accuracy, it can be assumed that the most recent values (i.e. those of the IPCC's Fifth and Fourth Assessment Reports) are the most accurate ones. Thus, there is a certain trade-off between comparability and consistency with the national inventories and the relevant IPCC guidelines, on the one hand, and accuracy, on the other.

### 3.9.6 Scope – production vs. consumption approach

Inventories and tools fundamentally use different methodologies to account emissions, which could be broadly classified into production- and consumption-based approaches. A production-based approach inherently relies on the 'territoriality' principle of the IPCC, and attempts to estimate GHG emissions on location based perspective, within the geopolitical or legal boundaries of a city, region or country. The use of a production-based approach to assess the contribution of urban areas to GHG emissions can lead to factual representation of emission sources and affirmative action. Urban areas will be able to reduce their emissions through creating disincentives for dirty economic activities that generate high levels of GHGs (e.g. heavy industry), and incentives for clean economic activities that generate much smaller emissions (e.g. high-tech industries). This situation can already be seen: many polluting and carbon-intensive manufacturing pro-cesses are no longer located in Europe or North America, but have been sited elsewhere in the world (UN-Habitat, 2011, p. 58). Consequently, the responsi-bility of successful production-oriented centres such as Beijing and Shanghai is exaggerated, while that of wealthy service-oriented cities, including many cities in North America and Europe, is underemphasized. But on principle of territori-ality, by virtue of the legal and administrative mandate vested within the urban areas, and the ease and responsibility of enforcement, production-based emission methodologies seem highly practical.

On the other hand, a consumption-based approach attempts to address the origin of emissions in a more comprehensive manner, allocating responsibility of emissions to those who are actual end users of products and services, rather than to those areas where they have been produced. This type of accounting system would result in a lower level of GHG emissions in developing countries or their cities (with a likely substantial reduction in the GHG emissions allocated to China and Chinese cities), and should – in theory – influence consumers in developed countries to assume responsibility for choosing the best strategies and policies to reduce emissions (Bastianoni *et al.*, 2004). It is opined that consumption-based mechanisms inherently have greater degrees of uncertainty (as there are many more systems to be incorporated in the final calculation, making them virtually impossible to be used for policy and decision-making in an administrative and hierarchical structure), but they do provide considerable insight into climate policy and mitigation, and should probably be used at least as a complementary indicator to help analyse and inform climate policy (Peters, 2008).

Either of the approaches – production or consumption – can be benchmarked against global needs to limit GHG emissions to prevent dangerous climate change. The best available estimates suggest that annual global GHG emissions need to be reduced from approximately 50 to 20 bt $CO_2e$ per year by 2050. With an esti-mated global population of 9 bt $CO_2e$ in 2050, this means that individual carbon footprints around the world will have to be at an average of less than 2.2 bt $CO_2e$ per year, which should be fairly available to all, irrespective of national and local (urban or rural) origins. While the role of cities in contributing to climate change has not been undermined, a lot of conceptual research supporting consumption-based approaches transfers the allocation of responsibility of GHG emissions to the wealthy nations, organizations or individuals, and relegates city governments of their responsibility, although they are actually the main institutions involved at the local governance level to take mitigation and adaptation actions for climate change, thus creating a wider science-policy or theory-application vacuum.

*In order to rationalize the definition, boundary and production/consumption approach, the concept of 'Scopes' was developed* (discussed in Sections 3.4 and 3.6). Here, we examine their significance from production and consumption perspectives. Under Scope 1, city-based attribution takes into account GHG emissions from all production within the boundaries of a city (World Bank, 2010, p. 21). Under Scope 2, city-based attribution takes into account GHG emissions from city consumption, even if the production of emissions falls outside the boundary of a city. This includes emissions such as those produced by a power plant located outside of a city but whose power is consumed within the city. Under Scope 3, upstream emissions of cities are counted. This includes aviation and maritime emissions, which can increase a city's per capita GHG emissions by as much as 20%, depending on the connectivity of city residents. Scope 3 emissions also include upstream emissions from food production, landfills and fossil fuel processing. These upstream Scope 3 emission sources are an important component of city GHG emissions. Ramaswami *et al.* (2008) demonstrate that Denver's emissions increase by 2.9 t $CO_2e$/capita when the emissions from food and cement are included.

But in practice, GHG emission inventories from urban areas that include Scope 3 emissions are very rare. And the extent to which these Scope 3 emissions (i.e. indirect or embodied emissions) are included is very arbitrary, and there is no agreement for a framework to compare emissions of this type between urban areas. If Scope 3 or embodied emissions are included, it is likely that the per capita emission of GHGs allocated to a city will increase significantly, particularly if the city is large, well developed, and with a predominance of service and commercial activities (Dhakal, 2008). In addition, it is almost impossible to compile a comprehensive inventory of Scope 3 emissions that takes into account all consumption types of each and every individual living in an urban area. It has been further reported that per capita emissions of GHGs by individuals, including those caused by the goods they consume and wastes they generate, vary by a factor of more than 1,000, depending on the circumstances into which they were born and their life chances and personal choices. Unsustainable levels of consumption, which drive the processes of production, are therefore crucial to understanding the contribution of urban areas to climate change (UN-Habitat, 2011, p. 57).

The research underpins that there is no widely accepted international standard or protocol for GHG monitoring at the local level. Yet, in the absence of any such standard, most of the developers of inventory, tools, methodologies, etc. have taken the IPCC methodology for national inventories as the guideline. Given that the IPCC guidelines have not been developed for the local level, tool developers have therefore made some adaptations to it. Bader and Bleischwitz (2009) analyse the international comparability and adherence of different tools to the prevailing methodologies, and reckon that most of them are consistent with the IPCC standard. But tools differ with regard to the allocation of indirect emissions due to electricity and heat, in defining the point of use and point of generation. If an inventory is to be consistent with the IPCC guidelines, only emissions at the point of generation should be taken into account.

According to Chavez and Ramaswami (2011), the lack of a standardized method for city-scale GHG emissions accounting to date has produced inconsistent accounting approaches for cities throughout the world. This inconsistency is seen both in the wide variation of inclusions in city-scale GHG emissions accounting in the peer-reviewed literature, and lack of explicit statements on what the unit of analysis is (i.e. for whom is the accounting being conducted: a household consumption or community-wide energy use?). That means whether it considers a sample or all residences, business and industries located within the geopolitical boundary, also termed the 'geographical-based approach'. At other times, GHG accounting appears to primarily address the consumption by households within a community – this is a subset of a full consumption-based approach, although many researchers are not explicit in such delineation. Finally, when GHG accounting addresses economic final consumption (i.e. households, government and capital expenditures within a community), this is termed full consumption-based accounting. Hence, the three emerging methods propounded for city-scale GHG emissions accounting are: geographic boundary-limited accounting, transboundary infrastructure supply chain (TBIS) footprinting, and consumption-based footprinting.

*Production-/geographic-based accounting*: Boundary-limited geographic approaches to GHG emissions accounting are those, which are largely considered 'production-based', even though they include GHG from fuel combustion by final consumption (i.e. in homes and personal vehicles). In other words, this method accounts for GHG emissions from all production activities within a unit's geopolitical boundary, although direct GHG emissions from end use of energy in households are also included.

*Consumption-based footprinting*: The consumption-based approach accounts for global GHG emissions resulting from economic final consumption (i.e. households, government and capital investments) within a unit (city), including GHG emissions in imports, but excludes GHG from the production of exports within the boundary. This method traces GHG emissions fully upstream, outside of the community boundary, accounting for all transboundary activities that serve economic final consumption in the community. Household consumption surveys are often used to assess the impact of only household consumption on GHG emissions (Weber and Matthews, 2008b).

Applications wherein city-scale IO tables are downscaled from national IO tables must be used to address other components of final consumption (government and capital investments). However, the accuracy of downscaled IO tables in representing material and energy flows in cities is questioned by Chavez and Ramaswami (2011). Consumption-based GHG emissions footprinting accounts for all (in-boundary and transboundary) emissions resulting from economic final consumption in the community, while the in-boundary commercial industrial activities exported elsewhere are excluded, even though these local activities generate jobs and may also be shaped by local regulations. The method is especially suited to educate households about the global nature of their consumption.

*TBIS footprint*: The TBIS GHG footprint method is an innovative method developed by Ramaswami *et al.* (2008) that recognizes that cities are not like large nations, in that energy use to provide essential infrastructures such as electricity often occurs outside the geographic boundary of the city (e.g. food production). The TBIS method therefore borrows the concept of Scopes used in corporate/business GHG accounting (described earlier in section 3.4) to account for essential transboundary infrastructures serving cities. The method can be thought of as an infrastructure-based supply chain footprint for cities, accounting for GHG emissions from buildings and infrastructures (e.g. residential, commercial and industrial) within the city (regarded as Scope 1) and transboundary electric power supply, transboundary transportation (e.g. road, air and freight), fuel supply, water supply, waste management and construction materials infrastructures serving cities (regarded as Scopes 2 and 3).

Thus, GHG emission footprints associated with cities seek to measure and allocate the in-boundary and transboundary GHG emissions associated with cities in a manner that provides rigorous data and informs policymaking. One way of describing in-boundary and transboundary GHG emissions is through the idea of Scopes. It may be seen that problems associated with accounting Scope 3 emissions are threefold:

1    There is a greater likelihood that, considering activities and transboundary infrastructures and supply chains associated with Scope 3 emissions, the overall emission burdens of a city are bound to escalate significantly.
2    The inventory for Scope 3 emissions considers selected supply chains, which are either less complex to quantify or for which data is easily available or could be adapted within the inventory parameters. Many other supply chains associated with trade and consumer items are virtually impossible to track at an urban scale.
3    Reliability and accuracy of data sets for Scope 2 and Scope 3 emissions is extremely weak.

In view of the above issues, pursuing inventories that require comprehensive data and scope-related discretion for the local governments in developing countries is seriously debatable. After all, these are already facing the daunting task of provisioning basic amenities to the urban poor at an affordable price against serious technical and financial limitations. Moreover, this bottom-up approach of preparing consistent and comparable GHG inventory for individual cities, and aggregating them to estimate the total emission footprint of all urban settlements within a country, which are about 7,900 in case of India (as per Census of India, 2011), the entire procedure becomes questionable. The case of transforming cities, the urbanizing ones, such as Class III to Class V, becomes even more paramount, as they are the ones that either have no urban local body in position or face the most techno-financial issues of governance, and at the same time require immediate interventions to alter their rapidly exploding cores, polluting environs and GHG-emitting activities.

### 3.9.7 Tiers/uncertainty in measurement

Some protocols and programmes define 'tiers' to indicate different levels of accuracy. Often, three tiers are given, whereby the tier 1 method is the least accurate and the tier 3 method is the most accurate. The IPCC recommends using tier 2 or 3 methods to calculate the key emission sources for a national inventory (IPCC, 2006).

- A tier 1 approach is the least accurate emission estimation, as it does not follow emission factors specific to the country where assessment is being conducted.
- A more accurate tier 2 approach requires that default emission factors are replaced by country-specific emission factors that take account of country-specific data. For instance, a country-specific emission factor for fuel combustion would take account of the average carbon content of the fuel, fuel quality, carbon oxidation factors and the state of technology development.
- A tier 3 approach would, in addition, take account of the age of the equipment used to burn the fuel, control technology, operating conditions, the fuel type used and combustion technology. Such an approach represents the most accurate emission quantification. For big plants, data on plant-specific $CO_2$ emissions are increasingly available. It is good practice to use the most disaggregated site- and technology-specific emission factors available. However, for many local territories, the use of a tier 3 approach might be too complex (IPCC, 2006). If a local authority has access to country and regional emission factors for key activities, then the regional emission factors should be preferred.

### 3.9.8 Metric/unit of measurement

Inventories and tools follow different metrics to represent GHGs, such as cumulative volume (in t $CO_2$e or Mt $CO_2$e), average per person (t $CO_2$e/capita) and as a function of economic output (t $CO_2$e/GDP). Chavez and Ramaswami (2011), while reviewing several methodologies, demonstrate that one single metric (e.g. GHG/person) will probably not be suitable to represent GHG emissions associated with cities. A combination of variables such as GHG per unit city residents plus city employees or the totality of economic output may all serve as potential metrics for defining a low-carbon city. In addition to aggregating citywide metrics, such as GHG/person or GHG/GRP, sector-specific efficiency and consumption measures are also useful.

It is argued that particular metrics will need to be ranked and weighted across cities. Efficiency benchmarks already existing in the literature could be expanded on. It is likely to take a combination of various metrics, together, to help define a low-carbon city both for rigor and for policy relevance (Zhou *et al.*, 2010). It is relevant especially for residential/community wide emissions, excluding industrial or high-commercial activities. Though providing emission in per capita units is helpful to highlight city-boundary issues, as most policymakers and the public can relate easily to which people are being counted (World Bank, 2010, p. 21), it has been argued that the measure of "GHG emissions per capita" is different from the measure of an individual's 'carbon footprint' – as the carbon footprint takes into account the overall implications of an individual's activities, including the purchasing of manufactured goods (Dodman, 2009), once again suggesting diverging results obtained from production and consumption perspectives.

### *3.9.9 Purpose and application*

It is pertinent to understand the objective for which the inventory or tool is going to be used, whether it is scientific or to guide public policy, inform regulation and provision by the government, or usher a behavioural change in the society. This accordingly determines *Input, Method* and *Output* such as a 'metric' in which the results are reported in inventory. The government of India underpins this vital link, and observes that annual GHG inventories will provide an opportunity to measure the impact of the steps taken to reduce carbon intensities. Also, reporting with large time lags does not provide timely information for policymaking. Thus, the need to create a mechanism that can estimate the GHG emissions on a regular basis, particularly in light of the goals that India has set for itself on reducing its emissions intensity. This is a daunting task, as uncertainty exists in terms of gaps in the activity data, quality of the activity data and non-availability of emission factors, etc. (Planning Commission, 2011, p. 25).

In the case of models and tools, it further needs to be underscored whether the tool aims for optimization (track past and present GHG emission trends and model performance to the optimal level) or is designed for simulation (generating alternative scenarios in future). Greater compatibility or interoperability of tools would render it easier to compare results and thus facilitate this process. The ultimate goal should be to reduce emissions (Bader and Bleischwitz, 2009). In this regard, the time, inputs and methods invested in compilation of a GHG inventory should be reduced to the minimum so that human resources can be used for the implementation of emission-reducing measures.

## 3.10 Need for spatial disaggregation of national emissions

The above discussion on major inconsistencies in inventories, tools and methodologies in terms of their input, method and output demonstrates a large gap in estimating GHG emissions emanating from all the urban areas within a country (see Table 3.4). In practice, emission estimates or inventories are prepared either through voluntary signing up with a group (C40, ICLEI, WRI, etc.) that provides technical know-how, or executed through a survey where a group collects all the activity data and independently arrives at the emission figures (as in ICLEI's South Asian Inventories for 53 cities). In both cases, the municipal or local government lacks the institutional mechanisms and the technical capacities and direct involvement of regular staff to estimate GHG emissions, formulate targets for mitigation, take corrective action and monitor results. There is a strong need to have an alternative method where the technical planning staff are able to correlate their planning proposals bearing several spatial causations and contributions, with their emission implications.

*Table 3.4* Accepted methodologies across various scales and gap therein

| Scale | Standard/protocol/tool/methodology | Year of inception |
|---|---|---|
| National | • IPCC Guidelines for National Greenhouse Gas Inventories 2006 | 1996 |

*Gap – no inventory or methodology to estimate GHG emissions emanating from all the urban areas within a nation*

| Scale | Standard/protocol/tool/methodology | Year of inception |
|---|---|---|
| Urban/city | • GRIP, $CO_2$, GROBBLIANZ, EMSIG | 2000–2004 |
| | • Bilan Carbone – the $CO_2$ Calculator | 2005–2008 |
| | • International Local Government GHG Emission Analysis Protocol (ICLEI) | 2009 |
| | • International Standard for Determining Greenhouse Gas Emissions for Cities (UNEP, UN-Habitat and World Bank) | 2010 |
| | • Global Protocol for Community-Scale Greenhouse Gas Emissions (WRI, ICLEI, C40) | 2012 |
| | • Peer-reviewed methodologies | |
| Industry/corporate | • Greenhouse Gas Protocol (WRI) | |
| | • The International Organization for Standardization (ISO) 14064 | |

Source: Author

Tracing the pathway from emission sources back through the production system ultimately leads to individual emission drivers (Peters, 2010). In many cases, emission drivers could be considered as individual consumers, but in a broader perspective an individual consumer may have little control over, for example, existing infrastructure, production systems or government policies. Emission drivers, therefore, need to be considered at several scales and in different contexts, covering, for example, individuals, households, companies and different levels of government (Peters *et al.*, 2009). Hence, the importance of scientific measurement of GHGs within multiple spatial scales could not be undermined. The management of GHG emissions and removals is a very nascent field of enquiry and practice, and is inherently interdisciplinary. Decisions at all scales, from personal to global, will need to consider the implications of actions on the atmosphere (Gillenwater and Pulles, 2011).

But for cities, there are several challenges owing to this very complexity of spatial scale. Local governments responsible for the governance of subnational regions do not have the same information sources as the national governments when compiling national inventories for the purpose of reporting under the UNFCCC. Records of the flow of energy and materials are typically most accurate at the national level, due to national governments having governance over imports and exports. The need to analyze GHG gas emissions at a local community level means that a combination of national and local area information is likely to be required in order to model emissions (IEAP, 2010). This suggests a scope for meso-methods such as disaggregation of national GHG data to estimate urban contribution.

This methodology has found favour with Planning Commission (now NITI Aayog, the premier national policymaking institution in India). It believes that this approach will have two major benefits. First, it will allow validation of the results of the estimation done through the top-down level (synthesis), thereby improving the accuracy and confidence of the estimates. Second, it will form the basis of emission trading programmes, voluntary disclosure programmes, carbon or energy taxes, and regulations and standards on energy efficiency and emissions that may be brought about in the future (Planning Commission, 2011, p. 26). Given that the workload associated with the collection of local data is relatively high, some inventories derive parts of their data from national statistics (i.e. data for a specific sector are broken down from the national GHG inventory). Scaling data from national GHG inventories can be particularly interesting for a city if the relevant data are difficult to obtain at the local level and are not expected to represent a great share of overall emissions. Data collection at the local level (bottom-up approach) guarantees a relatively accurate inventory. Cities therefore face a trade-off between compiling an inventory as accurate as possible, on the one hand, and limiting the time needed for the undertaking, on the other. Most local inventories are therefore based on a mix of bottom-up and top-down approaches (Bader and Bleischwitz, 2009).

In order to uphold credibility in methodology, the methods and approaches of GHG measurement and management should therefore comply with scientific quality criteria, but even so, with the quality criteria as set out in the UNFCCC decisions and guidelines, the resulting information should (or shall) be not only true, but also adhere to TCCCA (as defined in these guidelines). For this, we will need guidance that produces GHG data that are 'good enough' rather than 'the best', and are always fit for use in the policy and decision processes that they are designed for, as well as in scientific applications (Pulles, 2011). The above discussion informs that a methodology to account GHGs for an entire urban area of a country should have the following features:

- Follow a systems approach fundamentally based on a top-down approach, disaggregating national GHG emissions, prepared under the UNFCCC communication (hence, invariably consistent with IPCC methodologies, TCCCA framework and essentially production-based).
- Be able to validate downscaled emissions with a bottom-up approach of emission estimation or assumptions from individual entities/cities.
- Should be less rigorous, minimize excessive inputs and methods on compilation, and be more governance-/policy-focused.

Second, as evident, all inventories and methodologies to account for city GHG emissions are concerned with estimations only for an individual city (the local urban scale), fundamentally on the principle of emission factors and activity data. In doing so, these considerably fall short to answer the basic question on how do nation's urban areas collectively contribute to its carbon (GHG) footprint. For India, it raises the issue of whether it is appropriate to calculate emissions from

7,935 individual towns and cities, and aggregate or upscale to estimate urban India's contribution to national GHG emissions, or if it is imperative to downscale national GHGs into urban and rural components.

In this regard, it is worth exploring the evolving body of work in spatial distribution or disaggregation of emissions to a subnational scale (Sethi, 2014). Different studies use a variety of techniques, ranging from a grid-based approach transposing, distributing or assuming national or global emissions over regional spatial units through complex geographical information software/applications (Dao and van Woerden, 2009; Theloke *et al.*, 2009), to a rather aggregation-based method where emissions or energy use or expenditures from various point- and mobile-based sources are reclassified for an administrative/spatial unit-like county, such as in Rue du Can *et al.* (2008), VandeWeghe and Kennedy (2007) and Andrews (2008). For Sydney, Lenzen *et al.* (2004) disaggregated total primary energy use in 14 areas, and followed this up with a GHG emissions atlas of Australia at the postal district level (Dey *et al.*, 2007). A pioneering study used the Vulcan emissions atlas to compare transportation and building emissions in urban, peri-urban and rural counties of the US (Parshall *et al.*, 2009).

In the case of India, disaggreagation of nationally reported emissions using spatial analysis based on the production or location perspective of the most significantly contributing sector, namely energy, is established in Sethi (2015). Here, a data set of 454 thermal units reported by the Central Electricity Authority for 2011–2012 from over 100 coal, diesel, gas and liquid fuel-based gas turbine plants has been spatially located in real time, using the universally available (and verifiable) Web-based mapping tools Google Earth and Wikimapia. The information has been assessed on the urban–rural gradient, using the population-based census definition for class/hierarchy of towns and their location with respect to the urban boundary. The thermal power plants have eventually been mapped as within the urban area, on the periphery, in the transitional or rurban zone, or absolutely rural. The results present an array of emissions, across the urban hierarchy and location in space, underpinning how substantial GHG emissions are attributable to urban and urbanizing areas.

Due to greater availability of standard global data sets with high confidence and precision, disaggregation of emissions to different sub-spatial units has particularly found favour to influence studies capturing the role of urban areas and urbanization in climate change discourse (Satterthwaite, 2008; see also the famous GRIP methodology by Carney *et al.*, 2009, shown in Box 3.3).

---

### Box 3.3    The GRIP methodology

GRIP has three methods to estimate emissions from each emitting sector. The method that is applied is dependant on the level of data available in each region. The key calculation that runs throughout this methodology is:

**Emissions $R_{GX} = R_{XA} \times EF_{G.X}{}^{Y}$**

Where:

*R* is the region (say, European Union)

*X* is the activity under examination (measured or estimated)

*EF* is the emission factor

*G* is the greenhouse gas

*Y* is the region's nation

Essentially, the emission of (*G*) as in $CO_2$, $CH_4$ and $N_2O$, emanating from activity (*X*) in region (*R*) is equal to the level of activity (*X*) occurring in region (*R*) multiplied by the emission factor (*EF*) for GHG (*G*) for the activity (*X*) in the country (*Y*).

When a measured amount of the activity is not available, there needs to be a way of estimating the activity (*X*):

**activity $X_R = ((R_I \times R_H) / (N_I \times N_H)) \times N_X$**

Where:

*R* is the region

*N* is national

*X* is the activity under consideration

*I* is the indicator (e.g. GDP per household, expenditure on fuels, waste incinerated/landfilled/recycled in tons)

*H* is households

Essentially, the *estimated* level of activity of emissions source (*X*) in region (*R*) is equal to a regional value (*I*) multiplied by the emission factor (*EF*) for GHG (*G*) for the activity (*X*) multiplied by the national activity ($N_x$):

**activity $X_R = (R_I / N_I) \times N_X$**

Where:

*R* is the region

*N* is national

*X* is the activity under examination

*I* is the indicator (e.g. GVA (gross value added), population)

Essentially, the estimated level of activity of emissions source (*X*) in the region (*R*) is equal to a regional value (*I*) divided by the national indicator ($N_I$) multiplied by the national activity.

Source: Carney *et al.* (2009)

The above discourse suggests that methods based on spatial analysis could appropriately be used for carbon footprinting. Accordingly, spatial distribution or disaggregation of national emissions into urban and rural constituents from various sectors reported in India's national inventory for GHG emissions could be represented with the following equation. Applying this method, disaggregation of India's national GHG emissions for its urban contribution is performed and discussed in section 3.11.

The fundamental equation to disaggregate national emissions into urban and rural is thus:

$$GHG_{total} = GHG_{S1+S2+\cdots\cdots\cdots+Sn}$$
$$= [GHG_{S1.u1} + GHG_{S1.r1}] + [GHG_{S2.u2} + GHG_{S2.r2}] + \cdots\cdots\cdots$$
$$= [GHG_{Sn.un} + GHG_{Sn.rn}]$$
$$= [GHG_{Si.ui}] + [GHG_{Si.ri}]$$
$$= [GHG_{urban}] + [GHG_{rural}]$$

Where:

$S_i$ is the national GHG emissions from sector $i$

$u_i$ is the urban contribution to GHG emissions from an activity $i$ in sector $Si$

$r_i$ is the rural contribution to GHG emissions from an activity $i$ in sector $Si$

## 3.11 Urban India's role in contributing to GHG emissions

### 3.11.1 Quantitative assessment of sector contributions

Based on dissaggreagation frameworks such as Sethi (2015), Carney *et al.* (2009), etc., discussed above, and Barker *et al.* (2007) and Satterthwaite (2008), cited in *Cities and Climate Change* (UN-Habitat, 2011), Table 3.5 provides baseline urban contributions along with justifications to GHG contributions from the perspective of the location of activities that produced them. The sector contributions and their justifications, largely drawn from Sethi and Mohapatra (2013) on NATCOM II data sets submitted to the UNFCCC by the Indian Network for Climate Change Assessment (INCCA) (MoEF, 2012), are as follows:

*Electricity generation*: Throughout the world, fossil fuels are a major source of electricity. Globally, about two-thirds (67.1%) of 20,181 TWh electricity generated is sourced from thermal power plants (IEA, 2011), firing coal/peat, oil and petroleum gas as fuel, and thereby creating 28,999.4 Mt $CO_2$e emissions. The sector contributed to 26% of global GHGs in 2004 (IPCC, 2007) and 25% in 2010 (IPCC, 2014). In the case of India, there are over 100 thermal power plants generating 138,806.18 MW of energy (CEA, 2012) and 719.31 Mt $CO_2$e (MoEF, 2010). Fuel mix is heavily dependent upon coal (90%), followed by natural gas (8%) and oil (2%). As per the Census of India's broad definition of urban areas followed in India, discussed in section 1.7.3(a), most thermal plants would qualify to be in

*Table 3.5* Contribution of Indian cities to national GHG emissions by sector

| S. no. | Sector | Mt CO₂e 1994 | Mt CO₂e 2007 | Sector contribution in total Indian GHG emissions (%) | CAGR (%) | Justifications to GHG contribution from the perspective of the location of activities that produced them | Urban India baseline emissions Mt CO₂e 2007 | Urban India sectoral emission (%) | Sectoral component of total urban India emission (%) | Urban world baseline emission (% of total) |
|---|---|---|---|---|---|---|---|---|---|---|
| 1 | Electricity | 355.03 | 719.3 | 37.8 | 5.6 | Predominantly urban | 719.3 | 100 | 59.26 | 8.6–13.0 |
| 2 | Transport | 80.28 | 142 | 7.5 | 4.5 | 87% is road-based, 57.44% of total motor vehicles in the country owned in urban households | 70.98 | 50.0 | 5.85 | 7.9–9.2 |
| 3 | Residential* | 78.89 | 137.8 | 7.2 | 4.4 | Urban households form 33.29% of total Indian households | 45.88 | 33.3 | 3.78 | 4.7–5.5 |
| 4 | Other Energy+ | 78.93 | 100.9 | 5.3 | 1.9 | Petroleum refining is a peri-urban phenomenon | 33.85 | 33.6 | 2.79 | |
| 5 | Cement | 60.87 | 129.9 | 6.8 | 6 | Predominantly urban | 129.92 | 100.0 | 10.70 | |
| 6 | Iron and steel | 90.53 | 117.3 | 6.2 | 2 | Predominantly urban | 117.32 | 100.0 | 9.66 | |
| 7 | Other industry^ | 125.41 | 165.3 | 8.7 | 2.2 | Other metal and chemical is urban | 38.92 | 23.5 | 3.21 | 7.8–11.6 |
| 8 | Agriculture | 344.48 | 334.4 | 17.6 | −0.2 | Non-urban | 0 | 0.0 | 0.00 | 0 |
| 9 | Waste | 23.23 | 57.73 | 3.0 | 7.3 | Urban emissions considered in the assessment by INCAA | 57.73 | 100.0 | 4.76 | 1.5 |
| 10 | LULUCF | 14.29 | −177 | | | | 0 | 0.0 | 0.00 | |
| | Total | 1,228.5 | 1,728 | | 2.9 | | 1,213.9 | 70.3 | 100.00 | 30.5–40.8 |

Source: Adapted from Sethi and Mohapatra (2013); author's estimates based on data from MoEF (2012) and similar frameworks in UN-Habitat (2011), derived from Barker *et al.* (2007) and Satterthwaite (2008)

* Residential emissions also include commercial sector.

+ Other energy includes petroleum refining and solid fuel manufacturing, agriculture and fisheries, fugitive emissions.

^ Other industries comprising of pulp/paper, leather, textiles, food processing, mining and quarrying, and non-specific industries comprising of rubber, plastic, watches, clocks, transport equipment, furniture, etc.

*census towns*, cities and urban agglomerations, and practically the entire emission is attributable to urban areas. This is extremely different from the world scenario, where large fossil fuel power stations are considered to be located outside urban areas and estimated to contribute 8.6–13% of global emissions (Satterthwaite, 2008). Based on the national urban population of 377 million (Census of India, 2011), the average urban contribution of electricity in India is thus about 1.90 t $CO_2e$ per capita annually.

*Transport*: Cities prosper on mobility by means of air-, sea-, road- and river-based modes. Historically, urban centres all over the world were located on natural gateways, trade highways and sea routes. By the middle of the 20th century, all 15 million-plus cities were seaports. Globally, transportation contributed to 13% of global GHGs in 2004 (IPCC, 2007) and 14% in 2010 (IPCC, 2014). In the last few decades, motorized transport has become the preferred choice and the significant polluter – about three out of four energy-related GHG emissions came from road vehicles in 2004. In India, transport is responsible for 7.5% of the national emissions. It has grown at 4.5% annually since the 1994 assessment, to levels of 142.04 Mt $CO_2e$ in 2007 (see Table 3.5). Road transport constitutes for the bulk of emissions (87%). With a paucity of any national-level accounting or survey on annual motor trips in Indian cities, it is difficult to assess the urban component. Ownership of motor vehicles in the urban households is thereby used as a substitute indicator to assess their relative usage. As per Census of India (2011), 57.44%, or a little more than half of motor vehicles (including two- or three-wheelers, cars and light motor vehicles), are owned in urban households. Hence, urban areas contribute to about 71 Mt $CO_2e$ emissions.

The average individual emissions are very low: 0.18 t $CO_2e$/capita against other global cities, such as London (1.18), New York (1.47) and Los Angeles (4.74) (Kennedy *et al.*, 2009). Low per capita transport emissions in Indian cities can be attributed to the high use of non-motorized transport (NMT), such as cycles, rickshaws and walking trips, in many smaller cities. As per government estimates, NMT trips range from 36% to 53% of total trips across cities with populations varying from 0.5 to 8 million (MoUD, 2008). Sustained use of public transport is paramount to reduce emissions in this category of cities, as it ranges from 9% to 22% of total trips in sampled Indian cities in the above study. The case of the capital city has seen reduction in direct $CO_2$ emissions from public transport after the fuel switch to natural gas, and, as a Central Road Research Institute study showed, the Delhi Metro has saved 1,660,000 vehicle km and 2,275 tons of GHG emissions by 2007 (SARC, 2007) since its inception in 1998.

*Residential and commercial*: GHG emissions from residential and commercial buildings are caused by electricity use, lighting and illumination, space heating and cooling, direct emissions from cooking, embodied energy within building materials, etc. Globally, 10.6 bt $CO_2e$, or 8% of GHG emissions were emitted in this sector in 2005 (IPCC, 2007) and 6.4% in 2010 (IPCC, 2014). In India, the sector contributes to 139.51 Mt $CO_2e$, or 7.5% of national emissions. India's urban population resides in 110,139,853 households (Census of India, 2011), which is about one-third (33.29%) of the total households. As such, about 46.5 Mt $CO_2e$, or 3.78% of national emissions, is attributable to this sector in urban areas, with average individual contribution of 0.12 t CO2e/capita annually. This is close to the world scenario where residential and commercial activities in urban areas are estimated to contribute 4.7–5.5% of global emissions (Satterthwaite, 2008). In most cities worldwide, residential and commercial land use constitute the major half of the settlement. Emissions in this sector in India are on the rise and exhibit a large urban–rural divide (Gupta and Chandiwala, 2009; Parikh and Parikh, 2011). In terms of lifestyle differences across household expenditure classes, the urban top 10% accounts for emissions of 4,099 kg per capita per year, while the rural bottom 10% accounts for only 150 kg per capita per year (Parikh *et al.*, 2009). In the context of cities, it becomes imperative to assess these emissions with respect to factors shaping the built environment, namely city size (in terms of population and area), city form (population density) and geographical location.

*Industry*: Direct GHG emissions from the industrial sector are about 7.2 Gt $CO_2e$, and total emissions, including indirect emissions, were about 12 Gt $CO_2e$ in 2004, which constitutes 19% of global GHG emissions (IPCC, 2007), and 21% in 2010 (IPCC, 2014). These include manufacture of iron and steel, metal processing, chemicals and fertilizers, cement production, pulp and paper, and are responsible for direct emissions. Irrespective of the nature of industrial activity, generalization suggests that GHG contribution of industrial cities across the globe is substantially higher than cities with non-industrial activities as the primary economic activity. It should be noted that it is the energy-intensive 'heavy' or 'manufacturing' industries that cause emissions, while many cities that support an equally vibrant industrious workforce in cottage industries and finishing industries are less carbon-emitting. In India, cement, iron and steel, and other industries contributed to emissions of 129.92 Mt $CO_2e$, 117.32 Mt $CO_2e$ and 165.31 Mt $CO_2e$, respectively, in 2007 (MoEF, 2010) and collectively 412.55 Mt $CO_2e$, or 22% of national emissions. They have, respectively, grown at 6%, 2% and 2.2% per annum since the first assessment in 1994.

Most of the cement, iron and steel industries in India are low-tech and labour-intensive. Considering the modest definition of urban areas adopted in India, it would be inappropriate to qualify them functioning as rural entities. As such, 247.24 Mt $CO_2e$ can virtually be attributed to urban areas. On the same grounds, 38.92 Mt $CO_2e$ emitted from "other metals and chemicals" in the 'other industries' category has urban associations. Hence, a total of 286.16 Mt $CO_2e$ is produced in cities and forms a bulk of industrial emissions (69.4%), and 16.6% of the total nationwide emissions. This is almost double the 7.9–9.2% of GHGs allocated to cities worldwide (Satterthwaite, 2008). The per capita industrial energy consumption in urban areas thus arrives at 0.76 t $CO_2e$/annum.

*Waste*: Waste management in urban agglomerations becomes a critical issue of hygiene, sanitation and GHGs. Global GHG emissions from the waste sector are 1,300 Mt $CO_2e$, or 2.8% of global GHG emissions (IPCC, 2007). Indian emissions from the waste sector are of the order of 57.73 Mt $CO_2e$, forming 3% of national emissions (MoEF, 2010). The methodology accounts only urban-generated wastes, assuming that waste in rural areas decomposes in disaggregated and aerobic conditions, thus not emitting GHGs. Thereby, the entire 57.73 Mt $CO_2e$ is assigned to urban origins, with average generation of 0.15 t $CO_2e$/capita. India is still to achieve complete sanitation in urban areas, as only 75–81% of urban households have toilets, while the average solid waste generation from 41 Indian cities is 0.61 kg/day (ICLEI, 2009). By 2030, the annual generation and collection figures are likely to increase to 377 Mt and 295 Mt, respectively, thus widening the gap fourfold (Mckinsey, 2010). One of the reasons for this gap, as government studies suggest, is limited methane gas recovery at landfill sites, incineration, and waste-to-energy projects installed (CPCB, 2006).

*Land use, land use change and forestry (LULUCF)*: Globally, agriculture and forestry activities are allocated 13.5% and 17.4%, respectively, of GHG emission (IPCC, 2007), and their collective contribution was estimated as 24% in 2010 (IPCC, 2014). Seemingly rural-based, it is worth considering that LULUCF activities support the world's urban population with food grains, horticulture products, forest produce, non-timber forest produce and industrial raw materials that fuel the economy. Indian cities in the last decade have grown 44% from natural population growth, and 56% from rural–urban migration and the change in boundary definitions (MoUD, 2011). In the absence of any policy or regulatory mechanism to permit or limit movement of people across the state or local jurisdictions, the municipal bodies in India have no power to control migration. But considering the location of activities from where the GHG is produced, the LULUCF-associated emissions are essentially rural in origins. As per the INCCA report, the sector contributes to a net reduction of 177.03 Mt $CO_2e$ (MoEF, 2010). It further states that while areas under settlements/built-up lands increased by 0.01 mHa in 2006–2007, the conversion did not lead to any additional emissions, but a net removal of 0.04 Mt $CO_2e$. This requires appropriate scientific corroboration through recent satellite data, considering that land demand for various government projects/schemes during the XII Five Year Plan (i.e. 2007–2012) has been 0.3 mHa (Sethi, 2011).

According to certain estimates, the world urban population is expected to double, and the built-up area triple, by 2030 (Angel *et al.*, 2005). This will pose enormous challenges to mitigating capacities of agricultural lands and other global commons such as forests, parks and city greens, etc. There exists a vast potential to intensify greenery within the prevailing city boundaries. As per recent figures (FSI, 2011), green cover of New Delhi has increased from 26 sq km (1997) to about 296.20 sq km (2009) by the setting up of an eco joint force. Urban forestry and agriculture holds a multifold potential, as it enhances mitigation and assimilative capacities of the urban system, reduces the heat island effect, thus limiting the use of air conditioning in buildings and vehicles, and also diminishes the transportation emissions on account of haulage of food and vegetables into cities from villages. The small but incremental land use changes for green areas within a city, and its impact on enhancing mitigation and assimilation capacities as a whole, needs further empirical research.

The downscaled emissions outlined in Table 3.5 indicate urban emissions from all sectors as 1,213.9 Mt $CO_2e$, which is 70.3% of total national emissions.

Most of the carbon emissions (59.26%) are attributable to electricity generation, thereby reflecting that the menace of the carbon-intensive thermal plants in the power sector has a strong urban association, which otherwise is not even a subject under the purview of urban ministries, development authorities or local governments. This is followed by the cement, iron and steel industries, which form the backbone of the burgeoning construction and real estate industry, and collectively emit about one-fifth of the urban GHGs (20.36%). *The annual average emissions from urban areas thus amount to 3.11 t $CO_2e$/capita, double the national levels of 1.4 t $CO_2e$/capita. This comes as jolting evidence to the assumption that urban India does not contribute much to national GHGs, considering that hardly one-third of the population actually resides there.*

The disaggregation emission assessment is verified with ICLEI data for a sample of 41 cities assessed in 2008. According to this study, the total emissions of a city range from 0.14 and 0.21 Mt $CO_2e$ (Shimla, Haldia) on the lower side, to 7.36 and 9.33 Mt $CO_2e$ (Vishakhapatnam, Kolkata) on the higher end. It is observed that cities with large populations and industrial bases are the most GHG-emitting. The per capita city-level emissions range from 0.25 t $CO_2e$/capita (Asansol, Thiruvananthapuram) to 2.76 t $CO_2e$/capita (Jamshedpur).

### 3.11.2 Validation of downscaled emissions

An advanced analysis of the inventory (see Table 3.6) reveals skewed sampling of cities in the ICLEI inventory. Only 18 out of 53 million-plus cities (about 34%) and 23 out of 415 Class I cities (which have at least 100,000 population), but excluding million-plus (about 5.5%), are included. No sample has been selected from Class II to Class VI towns, which form a considerable bulk of urban settlements (94%), housing about one-third (i.e. 29.8%) of the urban population. Out of the three megacities enumerated in Census of India (2011), Greater Mumbai (18.4 million), Delhi Urban Area (16.3 million) and Kolkata (14.1 million), only the emissions of the Kolkata Municipal Area (4.5 million in 2001) have been accounted for. The average city emissions for Class I cities (1.28 Mt $CO_2e$) and million-plus cities (4.92 Mt $CO_2e$) derived from ICLEI data is used to estimate the overall urban emissions in that particular category (i.e. 531.20 and 260.76 Mt $CO_2e$, respectively). But the above estimate needs appropriation for the megacities and smaller urban areas (i.e. Class II to Class VI towns). For megacities, data compiled by the World Bank in the *Cities and Climate Change* report for the top 50 cities indicates that the estimated emissions for Greater Mumbai is 25 Mt $CO_2e$, while that of Delhi-National Capital Territory is 24 Mt $CO_2e$, and Kolkata is 16 Mt $CO_2e$. Thereby, collective emission load of the three Indian megacities is appropriated as 65 Mt $CO_2e$ instead of 15 Mt $CO_2e$, based on averages for million-plus cities inferred from ICLEI data (see Table 3.6 for all data compilations).

For almost 7,468 smaller cities (Class II to Class VI towns), currently there are neither data on number of towns in Census of India (2011), nor GHG emissions. But we do know from Census of India (2011) that about 70% of the towns in this category are under Class II and Class III, with populations between

Table 3.6 Validation of downscaled emissions across different city class

| Class definition | | (1) No. of cities Census 2011 | (2) Contribution to urban population | (3) ICLEI sample cities | (4) Average city emissions (Mt CO₂e) | (5) Average city population | (6) = (4) / (5) Average per capita emissions (t CO₂e) | (7) = (1) × (4) Urban emissions | (8) Appropriated emissions |
|---|---|---|---|---|---|---|---|---|---|
| Megacities | 10,000,000+ | 3 | 12.9% | 0 | | | | | 65.00* |
| Million+ cities | 100,000–9,999,999 | 53~ | 29.7% | 18 | 4.92 | 3.741 | 1.32 | 260.76 | 195.76 |
| Class I cities (excluding million+ or megacities) | 100,000–999,999 | 415 | 27.6% | 23 | 1.28 | 0.634 | 2.02 | 531.20 | 531.20 |
| Class II – Class VI census towns/cities | 5,000–99,999 | 7,468 | 29.8% | 0 | | 0 | | 0 | 358.45^ |
| Total | | 7,936 | 100% | 41 | | | | 791.96 | 1,150.41 |

Source: Author's estimates based on data from Census of India (2011): columns (1) and (2), ICLEI (2009): columns (3)–(6).

* Emissions appropriated for three megacities considering data from World Bank (2010).
~ Including three megacities: Greater Mumbai, Delhi NCT and Kolkata.
^ Based on the author's assumptions, explained in section 3.11.2.

20,000 and 99,999, gradually advancing on emission trajectories similar to their higher-ranking city group (i.e. Class I towns included in ICLEI data), such as Shimla (population 140,000, 0.14 Mt $CO_2$e) and Haldia (population 170,000, 0.21 Mt $CO_2$e). Assuming moderate emission loads of 0.06 Mt $CO_2$e, the contribution of Class II and Class III towns is about 313.65 Mt $CO_2$e (0.06 × 70% × 7,468). Similarly, for the remaining 30% of smaller cities (Class IV, V and VI), representing towns with populations less than 20,000, an emission load of 0.02 Mt $CO_2$e is assumed to contribute 44.80 Mt $CO_2$e (0.02 × 30% × 7,468). Hence, the contribution of all 7,468 smaller cities is 358.45 Mt $CO_2$e (see Figures 3.2 and 3.3 to see the GHG contribution from cities of different class sizes).

Collectively, the appropriated emissions for urban India corroborated from ICLEI assessments, World Bank data and suitable assumptions are of the magnitude of 1,150 Mt $CO_2$e. Comparing the above validation of national urban emissions with downscaled emissions in section 3.11.1, it may be concluded that by all reasonable means, urban India emissions range from 1,150 to 1,214 Mt

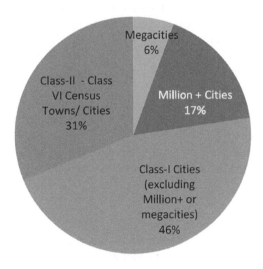

*Figure 3.2* Origin of GHG emissions from cities of different class size in India

*Figure 3.3* Spatial distribution of India's urban GHG emissions from cities of major class size

Note: Numbers in parenthesis denote the number of cities in that group and their average GHG emissions in Mt $CO_2$e. The Class II–VI cities group is a vast group of 7,468 cities, and extends beyond this visualization. Urban India's emission load/volume has been shown for representation purposes.

$CO_2$e, which is about 66.5–70.3% of the national GHG emissions. The annual per capita emissions from urban areas, obtained after downscaling techniques, average 3.11 t $CO_2$e/capita. While the national GHG emissions amount 1.4 t $CO_2$e/capita, considering that 68% of the total population lives in the countryside, the average rural emissions thus conclude to 0.66 t $CO_2$e/capita. It is noteworthy that this miniscule amount, about one-fifth of the urban counterpart, is grossly contradictory to the general notion prevalent in India that there is hardly any difference between national, urban and rural 'per capita' emissions.

---

**Box 3.4   Main empirical findings of India's urban GHG emissions**

- Urban India emissions range from 1,150 to 1,214 Mt $CO_2$e, which is about 66.5–70.3% of the total national GHG emissions.
- The quantum of India's urban GHG emissions is comparable to that of Japan (1,300 Mt $CO_2$e) and more than many Annex I or industrialized/developed countries, such as Germany (977 Mt $CO_2$e), Canada (742 Mt$CO_2$e), the UK (642 Mt $CO_2$e) and Australia (560.6 Mt $CO_2$e).
- Against the 32% urban population, Indian cities and towns contribute to about two-thirds of GHG emissions.
- GHG emission contribution: urban 3.11 t/capita vs. rural 0.66 t/capita.
- India's per capita emission is essentially low-carbon because of its minor rural contributions.
- Legal, technical, functional and financial capacities of the urban local bodies are limited to mitigate quantum emissions on their own.

---

In view of the above analysis, the research presents striking results for urban-based GHG emissions. It reveals that in the light of economic development, growing urbanization and GHG emissions, and inconsistencies in available methodologies and inventories, the role of urban areas in contributing to national GHG emissions cannot be undermined. Against the conventional view that India is largely a rural society, where urban areas house marginalized citizens, mostly living in slums unable to meet their energy needs, hardly contributing to GHG emissions, *this chapter brings forth that, in fact, Indian cities drastically contribute to global climate change, to an extent that while urban India is home to about one-third of the country's population, it contributes to at least two-thirds of national GHG emissions. Major sources of GHG emissions include electricity generation from thermal power plants, followed by the cement and iron and steel industries, to which cities at present have no or restricted mandate over their policy and regulation matters.* It is also imperative to further research on how urban emissions vary with their spatial features, such as size of the urban centre, city structure, urban form and activity/land use mix, as addressed in detail in Chapters 4 and 5. The above findings, along with inferences from subsequent chapters, will be valuable to influence *sustainable urbanization,* by limiting GHG emissions and ushering climate mitigation and co-benefits inclusive strategies in local urban governance.

---

## Basic questions

1   According to the UNFCCC protocol, what are the five criteria to ensure quality in GHG inventories?
2   Explain four main scientific methods for GHG quantification.
3   What is 'Scope' in certain standards and protocols that estimate GHGs, and how did it originate?
4   What are the three main types of city-/urban-based GHG emission methodologies? Explain any one of them.
5   Do all city-based GHG accounting methods conform to the UNFCCC principles of TCCCA and MRV? What are their methodical and practical limitations?
6   What is spatial disaggregation of national emissions, and why is it practically significant?

## Advanced questions

1   How is application of a GHG method subject to the scale at which assessment is to be performed? Substantiate it with some examples.
2   Discuss the 2006 IPCC Guidelines to estimate national GHG emissions and its relevance to subnational entities such as cities.
3   Discuss the methodologies to estimate emissions from corporations and businesses and its relevance to subnational entities such as cities.
4   List nine major inconsistencies in city GHG inventories, tools and methodologies. Discuss the two most crucial ones in detail.
5   How do product- or consumption-based approaches in accounting emissions of a territory influence its carbon footprint? Illustrate using two or three examples.

## Do it yourself exercises

1   Download any city-based GHG protocol/methodology/tool, etc. from the Internet and try to assess the annual carbon footprint of your city. For data not separately collected or classified by any agency, note down NC. Make sure you consult experts/officials as a lot of data exist with local agencies that are neither in the public domain nor commonly known. For data not readily available (though it may exist), note down NA. Are you able to effectively inventorize your city's GHG emissions? What are the problems you have faced? Calculate the combined percentage of NC and NA in your reporting.
2   Using the fundamental equation to disaggregate national emissions into urban and rural (given at the end of section 3.10), perform disaggregation analysis

for national urban emissions of your country. Employ standard data sets on GHGs, urbanization, households, etc. collected for the do it yourself exercises in Chapters 1 and 2. Draw reasonable assumptions and justifications in each sector. Evaluate and debate highly contributing and non-contributing sectors among your peer group. Using poster presentations, the exercise could be done within a class too.

## Suggested reading

### *Guidelines/protocols/methodologies/inventories*

1 Dhakal, S. (2004). *Urban energy use and greenhouse gas emissions in Asian megacities.* Kitakyushu, Japan: Institute for Global Environmental Strategies.
2 UNFCCC (2004). *Guidelines for the preparation of national communications by parties included in Annex I to the convention, part I: UNFCCC Reporting Guidelines on annual inventories.* Twenty-first session on Subsidiary Body for Scientific and Technological Advice, 6–14 December 2004, Buenos Aires. Retrieved from http://unfccc.int/resource/docs/2004/sbsta/08.pdf on 13 October 2014.
3 IPCC (2006). *2006 IPCC Guidelines for National Greenhouse Gas Inventories* [S. Eggleston, L. Buendia, K. Miwa, T. Ngara and K. Tanabe (Eds.)]. Kanagawa, Japan: Institute for Global Environmental Strategies. Retrieved from www.ipcc-nggip.iges. or.jp/public/2006gl/index.html on 12 September 2014.
4 ICLEI (2008). *Draft international local government GHG emissions analysis protocol, release version 1.0.* Retrieved from www.iclei.org/fileadmin/user_upload/documents/ Global/Progams/GHG/LGGHGEmissionsProtocol.pdf on 12 October 2014.
5 IEAP (2010). *International Local Government GHG Emissions Analysis Protocol (IEAP).* Retrieved from http://carbonn.org/fileadmin/user_upload/carbonn/Standards/ IEAP_October2010_color.pdf on 15 September 2013.
6 UNEP, UN-Habitat and World Bank (2010). *Draft International Standard for Determining Greenhouse Gas Emissions for Cities.* Presented at 5th World Urban Forum, Rio de Janeiro, Brazil, March 2010. Retrieved from www.unep.org/urban_ environment/PDFs/InternationalStd-GHG.pdf on 6 October 2014.
7 Chavez, A. and Ramaswami, A. (2011). Progress toward low carbon cities: Approaches for transboundary GHG emissions' footprinting, *Carbon Management, 2*(4), 471–82.

### *Review of methodologies/tools/inventories, etc.*

1 Bader, N. and Bleischwitz, R., (2009). Measuring urban greenhouse gas emissions: The challenge of comparability. *S.A.P.I.EN.S., 2*(3).
2 Kennedy, C., Steinberger, J., Gasson, B., Hansen, Y., Hillman, T., Havranek, M., Paraki, D., Phdungsilp, A., Ramaswami, A. and Mendez, G.V. (2009). Greenhouse gas emissions from global cities. *Environmental Science and Technology, 43*, 7297–302.
3 Dhakal, S. (2010). GHG emissions from urbanization and opportunities for urban carbon mitigation. *Current Opinion in Environmental Sustainability, 2*, 277–83.
4 Gillenwater, M. (2011). Filling a gap in climate change education and scholarship. *Greenhouse Gas Measurement & Management, 1*, 11–16. doi:10.3763/ ghgmm.2010.0012.
5 UN-Habitat (2011). *Cities and climate change: Global report on human settlements.* London/Washington, DC: Earthscan & UNCHS.

**Spatially relevant methods**

1  Lenzen, M., Dey, C. and Foran, B. (2004). Energy requirements of Sydney households. *Ecological Economics, 49*(3), 375–99.
2  Rue du Can, S., Wenzel, T. and Fischer, M. (2008). *Spatial disaggregation of CO$_2$ emissions for the state of California*. Berkeley, CA: Environmental Energy Technologies Division.
3  Druckman, A. and Jackson, T. (2009). The carbon footprint of UK households 1990–2004: A socio-economically disaggregated, quasi-multiregional input-output model. *Ecological Economics, 68*, 2066–77.
4  Parshall, L., Gurney, K., Hammer, S.A., Mendoza, D., Zhou, Y. and Geethakumar, S. (2009). Modeling energy consumption and CO$_2$ emissions at the urban scale: Methodological challenges and insights from the United States. *Energy Policy.* doi:10.1016/j.enpol.2009.07.006.
5  Peters, G.P. (2010). Carbon footprints and embodied carbon at multiple scales. *Current Opinion in Environmental Sustainability, 2*, 245–50.
6  Sethi, M. (2015). Location of greenhouse gases (GHG) emissions from thermal power plants in India along the urban-rural continuum. *Journal of Cleaner Production, 103*, 586–600. doi:10.1016/j.jclepro.2014.10.067.

# References

Andrews, C. (2008). Greenhouse gas emissions along the rural–urban gradient. *Journal of Environmental Planning and Management, 51*(6), 847–70.
Angel, S., Sheppard S.C. and Civco, D.L. (2005). *The dynamics of global urban expansion.* Washington, DC: World Bank.
Bader, N. and Bleischwitz, R., (2009). Measuring urban greenhouse gas emissions: The challenge of comparability. *S.A.P.I.EN.S., 2*(3).
Baiocchi, G., Minx, J. and Hubacek, K. (2010). The impact of social factors and consumer behavior on carbon dioxide emission in the United Kingdom: A regression based on input-output and geodemographic consumer segmentation data. *Journal of Industrial Ecology, 14*, 50–72.
Barker T., Bashmakov, I., Bernstein, L., Bogner, J.E., Bosch, P.R., Dave, R., Davidson, O.R., Fisher, B.S., Gupta, S., Halsnæs, K., Heij, G.J., Kahn Ribeiro, S., Kobayashi, S., Levine, M.D., Martino, D.L., Masera, O., Metz, B., Meyer, L.A., Nabuurs, G-J., Najam, A., Nakicenovic, N., Rogner, H-H., Roy, J., Sathaye, J., Schock, R., Shukla, P., Sims, R.EH., Smith, P., Tirpak, D.A., Urge-Vorsatz, D. and Zhou, D. (2007). *Technical summary. In Climate change 2007: Mitigation. Contribution of working group III to the fourth assessment report of the Intergovernmental Panel on Climate Change* [B. Metz, O.R. Davidson, P.R. Bosch, R. Dave and L.A. Meyer (Eds.)]. Cambridge/New York: Cambridge University Press.
Bastianoni, S., Pulselli, F. and Tiezzi, E. (2004). The problem of assigning responsibility for greenhouse gas emissions. *Ecological Economics, 49*(3), 253–7.
Brown, M., Southworth, F. and Sarzynski, A. (2008). Shrinking the carbon footprint of metropolitan America. Washington DC: Brookings Institution. Retrieved from www.brookings.edu/papers/2008/05_carbon_footprint_sarzynski.aspx on 22 September 2014.
Carney, S., Green, N., Wood, R. and Read, R. (2009). Greenhouse gas emissions inventories for eighteen European regions, EU CO2 80/50 Project Stage 1: Inventory formation. *The greenhouse gas regional inventory protocol (GRIP).* Manchester: Centre for Urban and Regional Ecology, School of Environment and Development, University of Manchester.
Census of India (2011). *Provisional Population Totals 2011,* Paper II, 2. New Delhi: Census of India.

124 *Role of cities: methods, tools & evidence*

CEA (2012) Renovation, modernisation and life extension of thermal power stations. *Quarterly Review Report (July–September 2012)*. Retrieved from www.cea.nic.in/ on 15 September 2013.

Chavez, A. and Ramaswami, A. (2011). Progress toward low carbon cities: Approaches for transboundary GHG emissions' footprinting, *Carbon Management, 2*(4), 471–82.

Churkina, G., Brown, D.G. and Keoleian, G. (2009). Carbon storage in human settlements: The conterminous United States. *Global Change Biology, 16*, 135–43.

CPCB (2006). *Assessment of status of municipal solid waste management in metro cities and state capitals*. New Delhi: Central Pollution Control Board.

Dao, H. and van Woerden, J. (2009). Population data for climate change analysis. In J.M. Guzman, G. Martine, G. McGranaghan, D. Schensul and C. Tacoli (Eds.), *Population dynamics and climate change* (pp. 218–38). New York/London: UNFPA & IIED.

Davis, S.J. and Caldeira, K. (2010). Consumption-based accounting of $CO_2$ emissions. *Proceedings of the National Academy of Sciences, 107*, 5687–92.

Dey, C., Berger, C., Foran, B., Foran, M., Joske, R., Lenzen, M. and Wood, R. (2007). An Australian environmental atlas: Household environmental pressure from consumption. In G. Birch, (Ed.), *Water, wind, art and debate: How environmental concerns impact on disciplinary research* (pp. 280–315). Sydney: Sydney University Press.

Dhakal, S. (2004). *Urban energy use and greenhouse gas emissions in Asian megacities*. Kitakyushu, Japan: Institute for Global Environmental Strategies.

Dhakal, S. (2008) Climate change and cities: The making of a climate friendly future. In P. Droege (Ed.), *Urban energy transition: From fossil fuels to renewable power* (pp. 173–92). Oxford: Elsevier Science.

Dhakal, S. (2010). GHG emissions from urbanization and opportunities for urban carbon mitigation. *Current Opinion in Environmental Sustainability, 2*, 277–83.

Dodman, D. (2009). Blaming cities for climate change? An analysis of urban greenhouse gas emissions inventories. *Environment and Urbanization, 21*(185). doi:10.1177/0956247809103016.

Dore, A.J., Vieno, M., Fournier, N., Weston, K.J., and Sutton, M.A. (2006). Development of a new wind-rose for the British Isles using radiosonde data, and application to an atmospheric transport model. *Quarterly Journal of the Royal Meteorological Society, 132* (621), 2769–2784.

Druckman, A. and Jackson, T. (2009). The carbon footprint of UK households 1990–2004: A socio-economically disaggregated, quasi-multiregional input-output model. *Ecological Economics, 68*, 2066–77.

Finkbeiner, M. (2009). Carbon footprinting: Opportunities and threats. *International Journal of Life Cycle Assessment, 14*, 91–4.

Forstall, R.L., Greene, R.P. and Pick, J.B. (2009). Which are the largest? Why lists of major urban areas vary so greatly. *Tijdschrift voor economische en sociale geografie, 100*(3), 277–97.

FSI (2011). *India Report*. New Delhi: Ministry of Environment and Forests, Government of India.

Gillenwater, M. (2011). Filling a gap in climate change education and scholarship. *Greenhouse Gas Measurement & Management, 1*, 11–16. doi:10.3763/ghgmm.2010.0012.

Gillenwater, M. (2015). Where can I find carbon management research? GHG Management Institute. Retrieved from https://ghginstitute.org/2015/10/26/where-can-i-find-carbon-management-research on 14 October 2015.

Gillenwater M. and Pulles, T. (2011). Welcome to greenhouse gas measurement and management. *Greenhouse Gas Measurement & Management, 1*, 3. doi:10.3763/ghgmm.2010.ED01.

Gupta, R. and Chandiwala, S. (2009). A critical and comparative evaluation of approaches and policies to measure, benchmark, reduce and manage $CO_2$ emissions from energy use in the existing building stock of developed and rapidly-developing countries: Case studies of UK, USA, and India. Paper presented at the 5th Urban Research Symposium, *Cities and climate change*: *Responding to an urgent agenda*, 28–30 June, Marseille, France.

Hertwich, E.G. (2005). Lifecycle approaches to sustainable consumption: A critical review. *Environmental Science and Technology*, *39*, 4673–84.

Hillman, T. and Ramaswami, A. (2010). Greenhouse gas emission footprints and energy use benchmarks for eight U.S. cities. *Environmental Science and Technology*, *44*, 1902–10.

Huang, Y.A., Lenzen, M., Weber, C.L., Murray, J. and Matthews, H.S. (2009). The role of input-output analysis for the screening of carbon footprints. *Economic Systems Research*, *21*, 217–42.

ICLEI (2008). *Draft international local government GHG emissions analysis protocol, release version 1.0.* Retrieved from www.iclei.org/fileadmin/user_upload/documents/Global/Progams/GHG/LGGHGEmissionsProtocol.pdf on 12 October 2014.

ICLEI (2009). *Energy and carbon emissions profiles of 54 South Asian cities.* New Delhi: International Council for Local Environmental Initiatives.

ICLEI (2014). *The five milestone process.* Retrieved from http://archive.iclei.org/index.php?id=810 on 2 January 2014.

IEA (2011). *CO₂ emissions from fuel combustion.* Paris: IEA/OECD.

IEAP (2010). *International Local Government GHG Emissions Analysis Protocol (IEAP).* Retrieved from http://carbonn.org/fileadmin/user_upload/carbonn/Standards/IEAP_October2010_color.pdf on 15 September 2013.

IPCC (2006). *2006 IPCC Guidelines for National Greenhouse Gas Inventories* [S. Eggleston, L. Buendia, K. Miwa, T. Ngara and K. Tanabe (Eds.)]. Kanagawa, Japan: Institute for Global Environmental Strategies. Retrieved from www.ipcc-nggip.iges.or.jp/public/2006gl/index.html on 12 September 2014.

IPCC (2007). *Climate change 2007: Mitigation. Contribution of working group III to the fourth assessment report of the Intergovernmental Panel on Climate Change.* Cambridge/New York: Cambridge University Press.

IPCC (2014). *Climate change 2014: Synthesis report. Contribution of working groups I, II and III to the fifth assessment report of the Intergovernmental Panel on Climate* [Core Writing Team, R.K. Pachauri and L.A. Meyer (Eds.)]. Geneva: IPCC.

Kates, R., Mayfield, M., Torrie, R. and Witcher, B. (1998) Methods for estimating greenhouse gases from local places. *Local Environment*, *3*(3), 279–98.

Kennedy, C., Steinberger, J., Gasson, B., Hansen, Y., Hillman, T., Havranek, M., Paraki, D., Phdungsilp, A., Ramaswami, A. and Mendez, G.V. (2009). Greenhouse gas emissions from global cities. *Environmental Science and Technology*, *43*, 7297–302.

Larsen, H.N. and Hertwich, E.G. (2009). The case for consumption-based accounting of greenhouse gas emissions to promote local climate action. *Environmental Science and Policy*, *12*, 791–8.

Lenzen, M. (2009). The path exchange method for hybrid LCA. *Environmental Science and Technology*, *43*, 8251–6.

Lenzen, M., Dey, C. and Foran, B. (2004). Energy requirements of Sydney households. *Ecological Economics*, *49*(3), 375–99.

Lenzen, M., Murray, J., Sack, F. and Wiedmann, T. (2007). Shared producer and consumer responsibility: Theory and practice. *Ecological Economics*, *61*, 27–42.

Lenzen, M. and Peters, G.M. (2010). How city dwellers affect their resource hinterland: A spatial impact study of Australian households. *Journal of Industrial Ecology*, *14*, 73–90.

Lin, T., Yunjun, Y. Xuemei, B., Feng, L. and Wang, J. (2013). Greenhouse gas emissions accounting of urban residential consumption: A household survey based approach. *PLOS ONE*, *8*(2). doi:10.1371/journal.pone.0055642.

Matthews, H.S., Hendrickson, C.T. and Weber, C.L. (2008). The importance of carbon footprint estimation boundaries. *Environmental Science & Technology*, *42*(16), 5839–42.

Mckinsey (2010). *India's urban awakening: Building inclusive cities sustaining economic growth*. New Delhi: Mckinsey Global Institute.

MoUD (2008). *Study on traffic and transportation policies and strategies in urban areas in India*. New Delhi: Ministry of Urban Development, Government of India.

MoUD (2011). *India's urban demographic transition: The 2011 census results-provisional*. New Delhi: JNNURM Directorate and National Institute of Urban Affairs, pp. 2–4

MoEF (2010). *India's GHG emissions profile*. New Delhi: Climate Modeling Forum, Ministry of Environment & Forests, Government of India.

MoEF (2012). *India: Greenhouse gas emissions 2007*. New Delhi: Indian Network for Climate Change Assessment, Ministry of Environment & Forests, Government of India.

Parikh, J. Panda, M., Ganesh-Kumar, A. and Singh, V. (2009). $CO_2$ emissions structure of Indian economy. *Energy*. Retrieved from www.irade.org on 18 June 2013.

Parikh, J. and Parikh, K. (2011). India's energy needs and low carbon options. *Energy*. Retrieved from www.irade.org on 18 June 2013.

Parshall, L., Gurney, K., Hammer, S.A., Mendoza, D., Zhou, Y. and Geethakumar, S. (2009). Modeling energy consumption and $CO_2$ emissions at the urban scale: Methodological challenges and insights from the United States. *Energy Policy*. doi:10.1016/ j.enpol.2009.07.006.

Peters, G.P. (2008). From production-based to consumption-based national emission inventories. *Ecological Economics*, *65*(1), 13–23.

Peters, G.P. (2010). Carbon footprints and embodied carbon at multiple scales. *Current Opinion in Environmental Sustainability*, *2*, 245–50.

Peters, G.P. and Hertwich, E.G. (2008a). Trading Kyoto: Nature reports. *Climate Change*, *2*, 40–1.

Peters, G.P. and Hertwich, E.G. (2008b): Post-Kyoto greenhouse gas inventories: Production versus consumption. *Climatic Change*, *86*, 51–66.

Peters, G.P., Marland, G., Hertwich, E.G., Saikku, L., Rautiainen, A. and Kauppi, P.E. (2009). Trade, transport, and sinks extend the carbon dioxide responsibility of countries. *Climatic Change*, *97*, 379–88.

Planning Commission (2011). *Interim Report of Expert Group on Low Carbon Strategies for Inclusive Growth*. New Delhi: Planning Commission, Government of India. Retrieved from http://planningcommission.nic.in/reports/genrep/Inter_Exp.pdf p25 on 19 November 2012.

Pulles, T. (2011). Greenhouse gas measurement and management: Why do we need this journal? (Editorial). *Greenhouse Gas Measurement & Management*, *1*, 4–6.

Ramaswami, A., Hillman, T., Janson, B. *et al.* (2008). A demand-centered, hybrid lifecycle methodology for city-scale GHG inventories. *Environmental Science & Technology*, *42*(17), 6455–61.

Rue du Can, S., Wenzel, T. and Fischer, M. (2008). *Spatial disaggregation of $CO_2$ emissions for the state of California*. Berkeley, CA: Environmental Energy Technologies Division.

SARC (2007) *6th Report: Local Governance*. New Delhi: Government of India.

Satterthwaite, D. (2007). *The transition to a predominantly urban world and its underpinnings*. Human Settlements Discussion Paper. London: IIED.

Satterthwaite, D. (2008). Cities' contribution to global warming: Notes on the allocation of greenhouse gas emissions. *Environment & Urbanization*, *20*(2), 539–49.

Sethi, M. (2011). Alternative perspective for land planning in today's context: Minimising compensation issues. *Indian Valuer*, *43*(4), 443–48.

Sethi, M. (2015). Location of greenhouse gases (GHG) emissions from thermal power plants in India along the urban–rural continuum. *Journal of Cleaner Production, 103,* 586–600. doi:10.1016/j.jclepro.2014.10.067.

Sethi, M. and Mohapatra, S. (2013). Governance framework to mitigate climate change: Challenges in urbanising India. In Huong Ha and Tek Nath Dhakal (Eds.), *Governance approaches to mitigation of and adaptation to climate change in Asia* (pp. 200–30). Hampshire, UK: Palgrave Macmillan.

Sovacool, B.K. and Brown, M.A. (2010). Twelve metropolitan carbon footprints: A preliminary comparative global assessment. *Energy Policy, 38* (9), 4856–4869.

Suh, S. and Nakamura, S. (2007). Five years in the area of input–output and hybrid LCA. *International Journal of Life Cycle Assessment, 12,* 351–2.

Theloke, J., Thiruchittampalam, B., Orlikova, S., Uzbasich, M. and Gauger, T. (2009). *Methodology development for the spatial distribution of the diffuse emissions in Europe.* Retrieved from http://prtr.ec.europa.eu/docs/Methodology_Air.pdf.pdf on 17 October 2014.

Tukker, A., Cohen, M.J., Hubacek, K. and Mont, O. (2010). The impacts of household consumption and options for change. *Journal of Industrial Ecology, 14,* 13–30.

UNEP, UN-Habitat and World Bank (2010). *Draft International Standard for Determining Greenhouse Gas Emissions for Cities.* Presented at 5th World Urban Forum, Rio de Janeiro, Brazil, March 2010. Retrieved from www.unep.org/urban_environment/PDFs/InternationalStd-GHG.pdf on 6 October 2014.

UNFCCC (2004). *Guidelines for the preparation of national communications by parties included in Annex I to the convention, part I: UNFCCC Reporting Guidelines on annual inventories.* Twenty-first session on Subsidiary Body for Scientific and Technological Advice, 6–14 December 2004, Buenos Aires. Retrieved from http://unfccc.int/resource/docs/2004/sbsta/08.pdf on 13 October 2014.

UN-Habitat (2011). *Cities and climate change: Global report on human settlements.* London/Washington, DC: Earthscan & UNCHS.

VandeWeghe, J.R. and Kennedy, C. (2007). A spatial analysis of residential greenhouse gas emissions in the Toronto Census Metropolitan Area. *Journal of Industrial Ecology, 11*(2), 133–44.

Weber, C.L. and Matthews, H.S. (2008a). Food-miles and the relative climate impacts of food choices in the United States. *Environmental Science and Technology, 42,* 3508–13.

Weber, C.L. and Matthews, H.S. (2008b). Quantifying the global and distributional aspects of American household carbon footprint. *Ecological Economics, 66,* 379–91.

Weber C.L. and Peters, G.P. (2009). Climate change policy and international trade: Policy considerations in the United States. *Energy Policy, 37,* 432–40.

Wiedmann, T. (2009). A review of recent multi-region input–output models used for consumption-based emissions and resource accounting. *Ecological Economics, 69,* 211–22.

Wiedmann, T.O., Lenzen, M. and Barrett, J.R. (2009). Companies on the scale: Comparing and benchmarking the sustainability performance of business. *Journal of Industrial Ecology, 13,* 361–82.

Williams, E.D., Weber, C.L. and Hawkins, T.R. (2009). Hybrid framework for managing uncertainty in life cycle inventories. *Journal of Industrial Ecology, 13,* 928–44.

World Bank (2010). *Cities and climate change: An urgent agenda,* Washington, DC: IRDC.

World Resources Institute (2002). *Designing a customized greenhouse gas calculation tool.* Retrieved from http://pdf.wri.org/GHGProtocol-Tools.pdf on 11 August 2014.

Zhou N, Price L and Ohshita S. (2010). A low carbon development guide for local government actions in China. Presented at *The US–China Workshop on Pathways Toward Low Carbon Cities.* Hong Kong, China, 13–14 December 2010.

# 4    Urban spatial parameters

*The GHGs of a city are embedded in the DNA of its urban spatial parameters.*

There are various urban factors that cause, influence or drive GHG emissions. Normally, studies into causations of environmental impacts and emissions have been understood through the IPAT equation, as a product of population or demographic factors, affluence of the society in question and the level of technology that is at their disposal, as discussed at length in section 1.7. The same principles are used to formulate inventories, tools and methodologies to account GHG emissions, by multiplying the activity data collated for the study unit (nation, region/ city, household, etc.) with the respective emission factors. But in the process, some other vital physical or spatial attributes that influence carbon throughput could be left out. For instance, in the case of cities, their geographical location and climatic conditions are basic attributes that influence its energy demand, and hence its GHG emissions, but are not studied directly in the inventories, tools, etc. This chapter studies the causal relationship, both conceptual and empirical, of urban spatial factors – drivers, parameters, indicators, etc. – to GHG emissions, beyond the conventional or non-spatial parameters.

## 4.1 Causations and drivers of GHG emissions

There are several causations and drivers attributed in literature – both normative and empirical – that influence GHG emission. These could be broadly classified as population-, affluence-, technology- and spatially-driven for each sector, as enumerated in Table 4.1.

## 4.2 Non-spatial parameters and urban spatial parameters

Population, economy (GDP/affluence), technology or energy sources are the main non-spatial parameters or drivers that influence GHG emissions. These include sub-parameters such as residential or commercial electricity consumption, local $CO_2/NO_2$ levels for fuel emissions in ambient air, Bharat Standard (BS) akin to the Euro standard for vehicles (BSI, BSII, BSIII, BSIV, etc.). In addition, these include the use of diesel generation sets, coal-based power plants, transport fuel

Table 4.1 Causations and drivers of GHG emissions identified in literature

| Contributing sectors | Causations and drivers of GHG emissions | | | |
| --- | --- | --- | --- | --- |
| | Population | Affluence | Technology | Spatial |
| **Energy supply** | • Amount of electricity consumed (Kennedy et al., 2009b)<br>• Source of energy (Baldasano et al., 1999) | | • GHG intensity of fuel source (Kennedy et al., 2009b)<br>• Industrial activity (Kennedy et al., 2009b)<br>• GHG intensity of fuel source (Sugar et al., 2012) | • Access (proximity) to energy source – hydropower or coal seams (Kennedy et al., 2009b)<br>• Location in relation to natural resources (UN-Habitat, 2011, p. 52) |
| **Transport** | • Automobile use (Kennedy et al., 2009b) | • Vehicle fuel economy (Kennedy et al., 2009b)<br>• Car ownership (Dhakal, 2004)<br>• Automobile use (Lankao, 2007)<br>• Private car ownership (Dodman, 2009) | • Use of fossil fuels (Kennedy et al., 2009b)<br>• Quality of public transit/modal split (Kennedy et al., 2009b)<br>• Nature of transportation system – modal split, non-motorized transport (Dhakal, 2004)<br>• Role of alternative fuels (Dhakal, 2004)<br>• Energy efficiencies of key technologies (Dhakal, 2004) | • Compact development and mixed land use (UN-Habitat, 2011, p. 54)<br>• Spatial structure (Bertaud et al., 2009)<br>• Urban morphology or density (Dodman, 2009)<br>• Urban form/density (Kennedy et al., 2009b)<br>• Land use planning (Kennedy et al., 2009b)<br>• Gateway status (Kennedy et al., 2009b)<br>• Compactness of urban settlements (Dhakal, 2004)<br>• Urban spatial structure and urban functions (Dhakal, 2004)<br>• Availability of road and railroad infrastructure (Dhakal, 2004)<br>• Gateway status (Sugar et al., 2012)<br>• Urban form (Sugar et al., 2012)<br>• Location of population and economic activities (Lankao, 2007)<br>• Scale of the city (Brown et al., 2008)<br>• Density (Satterthwaite, 1999)<br>• Development pattern (Brown et al., 2008)<br>• Urban structure (Baldasano et al., 1999) |

*(continued)*

Table 4.1 (continued)

*Causations and drivers of GHG emissions*

| Contributing sectors | Population | Affluence | Technology | Spatial |
|---|---|---|---|---|
|  |  |  |  | • Neighbourhood density (Gottdiener and Budd, 2005)<br>• Density and gasoline use percapita (Newman and Kenworthy, 1989)<br>• Scale of the city (VandeWeghe and Kennedy, 2007) |
| **Residential and commercial buildings** |  | • Average HH income (Kennedy *et al.*, 2009b)<br>• House size/quality of building envelope (Kennedy *et al.*, 2009b)<br>• Income level and lifestyle (Dhakal, 2004) | • Energy efficiencies of key technologies (Dhakal, 2004)<br>• Building technologies and floor space use for air conditioners, district heating, cooling, insulation and building energy systems define energy use (Dhakal, 2004) | • Urban sprawl (Satterthwaite, 2011)<br>• Climatic situation and altitude (UN-Habitat, 2011, p. 52)<br>• Geographic location (World Bank, 2010, p. 31)<br>• Home heating and temperature (Glaeser and Kahn, 2008)<br>• Heating degree days dependent upon local climate (Kennedy *et al.*, 2009b)<br>• Climate factors affect energy use and creation of urban heat islands (Dhakal, 2004)<br>• Compactness of urban settlements (Dhakal, 2004)<br>• Heating degree days dependent upon local climate (Sugar *et al.*, 2012)<br>• Weather – latitude and climate type (Brown *et al.*, 2008)<br>• Scale of the city (Brown *et al.*, 2008)<br>• Climate (Baldasano *et al.*, 1999)<br>• Compact development and mixed land use (UN-Habitat, 2011, p. 54)<br>• Density (Satterthwaite, 1999) |

| | | | |
|---|---|---|---|
| **Industry** | • Urban economy (UN-Habitat, 2011, p. 57)<br>• City economy (Baldasano et al., 1999) | • GHG intensity of fuel source (Kennedy et al., 2009b)<br>• Efficiency of industrial processes (Dhakal, 2004)<br>• Energy intensity (Satterthwaite, 2009) | • Urban functions – energy use in industrial and commercial cities (Dhakal, 2004)<br>• Location of industrial processes with respect to city borders (Dhakal, 2004) |
| **Agriculture** | | | • Carbon sequestrization potential, as sink dependent upon its land use/location is under debate (Kennedy et al., 2009b, citing IPCC Guidelines, 2006)<br>• Carbon sequestration by soils and plants, urban heat island (Jabareen, 2006) |
| **Forestry** | | | • Carbon sequestrization potential, as sink dependent upon its land use/location is under debate (Kennedy et al., 2009b, citing IPCC Guidelines, 2006)<br>• Deforestation, carbon sequestration by soils and plants, urban heat island (Jabareen, 2006) |
| **Waste management** | • Patterns of consumption and waste generation (Dodman, 2009)<br>• Waste generation (Lankao, 2007) | • Management of waste (Dodman, 2009)<br>• Methane recovery factor (Kennedy et al., 2009b)<br>• Waste management technologies (Dhakal, 2004) | |
| **Total emissions** | • Individual consumption patterns (Satterthwaite, 2011)<br>• Consumption of city's individuals, institutions and business (Satterthwaite, 2009) | | • Spatial location (Brown et al., 2008)<br>• Spatial location – production vs. consumption perspective (Satterthwaite, 2009) |

Note: Sector classifications follow Barker et al., (2007)

consumption, average trip lengths, use of metro as public transit, use of non-motorized transport, use of refuse-derived fuel (RDF) technology or incinerators in waste treatment, etc. Most of these are activity- and technology-driven, while the population only imputes a scalar effect on it.

Apart from studying parameters of population, affluence and technology that influence GHG emissions, in many cases there is another agent at play, which is generally assumed as a given or constant in this relationship – the spatial characteristics that defines a settlement. It is difficult to perceive in normal circumstances, but suppose the criterion of energy available from different sources, such as thermal power plants, nuclear energy or hydro energy, or emissions from industrial activities or high consumption from affluence in a city that cause additional burdens, are excluded from the equation, then we are left with the bare spatial determinants. These define a city's need to provide basic energy needs for residential consumption, heating or air conditioning/cooling and to commute, which is determined by its geographical location, local climate or gets locked up collectively through its spatial structure, density, built form, road pattern and land use, and eventually controlled by its size.

It is significant to disaggregate the activity for urban spatial parameters that define the quantum or scale of activity. The conventional knowledge or inventories while accounting GHG emissions considers urban spatial determinants as a given. It does not question the causation of the activity owing to spatial or physical characteristics and their variations. Though this may arguably seem correct, as normally one may not significantly articulate the physical determinants, such as geophysical conditions (including local climate), spatial structure, built form, scale, etc., to a large extent the manner in which they influence activities, and hence GHG emissions, could not be undermined. Conceptually, the relationship can be clearly understood with the support of the diagram shown in Figure 4.1.

In a process, say city transportation, the overall emissions could be estimated as a product of the activity and the emission factor characteristic of that process. Similarly, if the entire urban context, a municipality, city or city-region is considered as the basis of causations to the all metabolisms or multifarious activities that are responsible for emissions (or simply consider it to be the 'activity'), then the activity could be represented with the following equation:

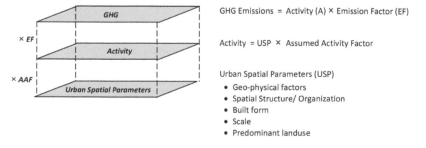

GHG Emissions = Activity (A) × Emission Factor (EF)

Activity = USP × Assumed Activity Factor

Urban Spatial Parameters (USP)
- Geo-physical factors
- Spatial Structure/ Organization
- Built form
- Scale
- Predominant landuse

*Figure 4.1* The relationship between GHGs, activity and urban spatial parameters

activity $(A)$ = urban spatial parameters $(USP)$ × assumed activity factor $(AAF)$

Where:

*USP* or urban spatial parameters is essentially the blueprint of an urban entity, defined by all its spatial attributes, such as geographical characteristics (including latitude, altitude, land form, gateway status and climate type), scale or size of the settlement, the spatial structure (including spatial organization or network pattern), built form and land use (including areas under greens, forests, roads, etc.)

*AAF* is the assumed constant of proportionality particular to that activity in the functional relationship

Various urban spatial parameters, as defined in different conceptual and empirical discourse, are reported from sections 4.3 to 4.8.

## 4.3 Geophysical conditions

Various aspects of geography affect the contribution of urban areas to climate change. These can be broadly categorized as climatic situation and altitude (Dhakal, 2010; Kennedy *et al.*, 2009b; UN-Habitat, 2011, p. 52). The climatic situation of any given urban area affects the energy demands for heating and cooling. High-latitude locations have longer hours of darkness in the winter, requiring additional energy consumption for lighting. High-latitude locations are also colder in winter, with additional heating requirements. Both space and water require heating, and in many countries the heating of water is a major consumer of energy in residential households. Heating requirements are usually met through the direct burning of an energy source such as coal, oil or natural gas. In contrast, space cooling through air conditioning is normally powered by electricity. Urban areas in warmer locations therefore have an emissions profile strongly influenced by the energy source used to generate this (UN-Habitat, 2011, p. 52).

In the US, the consumption of fuel oil and natural gas by households is determined primarily by climate. There is a very strong negative correlation between home heating-related emissions and lower temperatures in January – a factor that is itself determined by geographical location (Glaeser and Kahn, 2008). In contrast, many locations in the US with particularly hot summers (higher temperatures in July) have higher electricity consumption associated with space cooling. Solely taking these issues into account, it has been noted that areas with moderate temperatures have lower emissions and lower associated expenditure on energy. Although comparable studies have not been undertaken elsewhere, it is likely that this pattern is replicated on a global level, with areas experiencing very hot or cold climatic conditions requiring a greater use of energy for the cooling or heating of residential and commercial buildings (UN-Habitat, 2011, p. 52).

In many climate studies, geophysical conditions are defined by the indicator *heating degree days*, which determines the GHG emissions from a city. The consumption of fuels for heating and industry (i.e. stationary combustion in all

Table 4.2 Causation, indicators, affecting sectors and evidence of urban spatial parameters

| Parameter | Indicator | Relationship | Spatial causations | | |
|---|---|---|---|---|---|
| | | | Affecting sectors | Evidence | Source |
| *Geophysical conditions (explained in section 4.3)* | | | | | |
| Climatic situation and altitude | Climate type | Places with very hot or cold climatic conditions require a greater use of energy for the cooling or heating of residential and commercial buildings. Areas with moderate temperatures have lower emissions and lower associated expenditure on energy. | Residential and commercial buildings | No | UN-Habitat (2011, p. 52) |
| Climate | Temperature | There is a very strong negative correlation between home heating-related emissions and lower temperatures in January – a factor that is itself determined by geographical location. | | 66 major metropolitan areas within the US | Glaeser and Kahn (2008) |
| Geographic location | Heating and/or air conditioning requirements | Energy needs and GHG emissions from the residential sector are predominantly an outcome of essential district heating and/ or air conditioning requirements of the local people, which are in turn influenced by the settlement's geographic location. | Residential buildings | No | World Bank (2010, p. 31) |
| Climate | General reference to latitude as north and climatic regions such as temperate climate and urban heat island | Climate factors, especially excessive heat and cold climate conditions, directly affect energy use due to the greater demand for heating or cooling services. Cities in North Asia such as Beijing require more energy for space heating than do cities with temperate climates. Some cities in Asia, such as Tokyo and Seoul, also suffer from urban heat island (this is a phenomenon in which the core urban temperature is a few degrees higher than that in the suburbs; this creates | Energy use in residential and commercial buildings | Four Asian megacities: Beijing, Shanghai, Tokyo and Seoul | Dhakal (2004) |

| Parameter | Indicator | Relationship | Affecting sectors | Evidence | Source |
|---|---|---|---|---|---|
| | | hotspots in cities), where concentrated urban energy use is a major factor responsible for exacerbating urban warming. This sometimes triggers a vicious cycle in summer, where increasing the use of cooling devices contributes towards increasing the heat island effect. | | | |
| Climate | Heating degree days | Heating and industrial energy consumption for all global cities is closely related to climate. However, some cities have energy consumption exceeding that expected from the correlation to heating degree days. This is likely representative of an active industrial sector as in Shanghai. | Energy use in residential and commercial buildings | Chinese cities (selected) – Shanghai, Beijing and Tianjin | Sugar *et al.* (2012) |
| Weather | Latitudes, climate type loosely defined by thermal mildness and proximity to coast | Other factors are important, such as the fuels used to generate electricity, electricity prices and weather. Weather unmistakably plays a role in residential footprints. High-emitting metro areas often concentrate in climates that demand both significant cooling and heating, such as in the eastern mid-latitude states. In contrast, the 10 metro areas with the smallest per capita residential footprints are all located along the West Coast, with its milder climate. | Residential buildings | 100 largest metropolitan areas of the US | Brown *et al.* (2008) |

*Regional connectivity (explained in section 4.4)*

| Parameter | Indicator | Relationship | Affecting sectors | Evidence | Source |
|---|---|---|---|---|---|
| Location gateway | Air and sea linkages | The location of a city often determines its status as a gateway, thereby explaining emissions arising from airplanes and shipping. | Transport | 10 global cities, including Bangkok, Barcelona, Cape Town, Denver, Geneva, London, Los Angeles, New York, Prague and Toronto | Kennedy *et al.* (2009b) |

*Spatial causations*

(continued)

Table 4.2 (continued)

| Parameter | Indicator | Relationship | Affecting sectors | Evidence | Source |
|---|---|---|---|---|---|
| *Regional connectivity (explained in section 4.4)* | | | | | |
| Gateway | Aviation and marine activities | Cities are distinguished by the level of emissions from aviation and marine activity. Shanghai's per capita emissions from aviation and marine activity are the highest of the three cities, indicating that it is the most active hub of international economic activity and trade. Beijing's busy international airport is reflected in its high level of aviation emissions, while Tianjin's marine activity is more GHG-intensive than its aviation activity. | Transport | Chinese Cities (selected) – Shanghai, Beijing and Tianjin | Sugar *et al.* (2012) |
| Location within energy resource region | Proximity to energy source | Natural resources influence the fuels that are used for energy generation and, hence, the levels of GHG emissions, considering the factor of transportation costs. Access to hydropower, as in the cases of Geneva and Toronto, substantially reduces the intensity of emissions from these cities. Prague, on the other hand, lies close to some of the thickest coal seams in Europe. | | Rio de Janeiro and Sãu Paulo have low levels of emissions due to the availability of hydroelectric power from large rivers | UN-Habitat (2011, p. 52) |
| *Settlement scale (explained in section 4.5)* | | | | | |
| Scale | Population size | As can be seen from the wide variations in GHG emissions from countries around the world, population size in itself is not a major driver of global warming. It is the population growth rate, household size and consumer behavior that drives global warming. | Residential and industrial | Low-, lower-middle-, upper-middle- and high-income countries | UN-Habitat (2011, p. 53) |

| Parameter | Indicator | Relationship | Affecting sectors | Evidence | Source |
|---|---|---|---|---|---|
| Scale | Areal size (distance from city centre) | As the distance from the central core increases, private motor vehicle emissions begin to dominate the total emissions. | Transport | Toronto Census Metropolitan Area | VandeWeghe and Kennedy (2007) |
| Scale | Land area | Urban land areas are growing faster than ever because of a combination of absolute increases in numbers of people with a decreasing average density. This tendency towards declining density, combined with unprecedented absolute increases in the urban population, could greatly expand the land area of cities in the future. | Land use | All urban areas with populations more than 100,000 | Angel et al. (2005) |
| Scale | Population size | Large metropolitan areas offer greater energy- and carbon-efficiency than non-metropolitan areas. Despite housing two-thirds of the nation's population and three-quarters of its economic activity, the nation's 100 largest metropolitan areas emitted just 56% of US carbon emissions from highway transportation and residential buildings in 2005. | Transport and residential buildings | 100 largest metropolitan areas of the US | Brown et al. (2008) |

*Urban form (explained in section 4.6)*

| Parameter | Indicator | Relationship | Affecting sectors | Evidence | Source |
|---|---|---|---|---|---|
| Urban form | Compact development and mixed use | Spatially compact and mixed use urban developments have several benefits in terms of GHG emissions, which includes reduced costs for heating and cooling resulting from smaller homes and shared walls in multi-unit dwellings, lesser line losses related to electricity transmission and distribution, along with the possibility of using micro-grids at local level, reduced average daily vehicle kilometres travelled in freight deliveries and by private motor vehicles per capita. | Residential and commercial buildings and transport | No | UN-Habitat (2011, p. 54) |

(continued)

Table 4.2 (continued)

| Parameter | Indicator | Relationship | Affecting sectors | Evidence | Source |
|---|---|---|---|---|---|
| *Urban form (explained in section 4.6)* | | | | | |
| Form | Density | Population density increases accessibility to such destinations as stores, employment centres and theatres. | Transport | 13 American, seven Canadian, six Australian, 11 European and nine Asian cities | Newman and Kenworthy (1999) |
| Form | Density | The high concentrations of people and economic activities in urban areas can lead to economies of scale, proximity and agglomeration – all of which can have a positive impact upon energy use and associated emissions, while the proximity of homes and businesses can encourage walking, cycling and the use of mass transport in place of private motor vehicles. | Residential and commercial buildings and transport | No | Satterthwaite (1999) |
| Density | Neighbourhood density | Doubling of average neighbourhood density is associated with a decrease of 20–40% in per household vehicle use with a corresponding decline in emissions. | Transport | No | Gottdiener and Budd (2005) |
| Density | Gasoline use per capita | Gasoline use per capita declines with urban density, although this relationship weakens once GDP per capita is brought into consideration. | Transport | 13 global cities | Newman and Kenworthy (1989) |
| Urban form and horizontal sprawl | Density | Low-density suburban development in Toronto is 2 to 2.5 times more energy- and GHG-intensive than high-density urban core development on a per capita basis. | Three major elements of urban development are considered: construction materials for infrastructure (including residential dwellings, utilities and roads), building operations and transportation (private automobiles and public transit) | Toronto | Norman *et al.* (2006) |

| Parameter | Indicator | Relationship | Affecting sectors | Evidence | Source |
|---|---|---|---|---|---|
| Density | Energy use | Density may also affect household energy consumption. More compact housing uses less energy for heating. For example, households in the US living in single-family detached housing consume 35% more energy for heating and 21% more for cooling than comparable households in other forms of housing. | Residential buildings | No | UN-Habitat (2011, p. 55) |
| Compact form | Energy use | The compactness of urban settlements influences the demand for energy for transportation and for other areas such as district heating and cooling using co-generation systems. Urban sprawl, in which low-density suburbs depend on lengthy distribution systems, undermines efficient energy use. | Residential and commercial sector and transport | Four Asian megacities: Beijing, Shanghai, Tokyo and Seoul | Dhakal (2004) |
| Urban form | Density | Urban form also has a strong bearing on urban metabolism. As previous researchers have shown, transportation energy use is inversely correlated with urban population density. The analysis of the 10 cities here shows that such a relationship also holds for GHG emissions. | Transport | 10 global cities, including Bangkok, Barcelona, Cape Town, Denver, Geneva, London, Los Angeles, New York, Prague and Toronto | Kennedy et al. (2009b) |
| Urban form | Population density | The emissions from transportation are indicative of urban form, showing an inversely proportional relationship between transportation emissions and population density. The three Chinese cities are densely populated with low transport emissions. The Chinese cities have transportation emissions comparable to European cities with extensive public transport networks. | Transport | Chinese cities (selected) – Shanghai, Beijing and Tianjin | Sugar et al. (2012) |

(continued)

Table 4.2 (continued)

| Parameter | Indicator | Relationship | Affecting sectors | Evidence | Source |
|---|---|---|---|---|---|
| *Urban form (explained in section 4.6)* | | | | | |
| Urban sprawl | Dependence on private modes | The shift of residences from the city to the suburbs and beyond that these improvements brought was in part driven by the possibilities for those who moved to get more space and escape urban pollution and congestion. But the urban sprawl that this often produced brought with it a set of health problems, and generally a greater dependence on private automobile use, and so higher GHG emissions. | Transport | General – no evidence | Satterthwaite (2011) |
| Development pattern | Population density and public modes of transport | Development patterns and rail transit play an important role in determining carbon emissions. Density, concentration of development and rail transit all tend to be higher in metro areas with small per capita footprints. Much of what appears as regional variation may be attributed to these spatial factors. Dense metro areas such as New York, Los Angeles and San Francisco stand out for having the smallest transportation and residential footprints. Alternatively, low-density metro areas such as Nashville and Oklahoma City predominate in the 10 largest per capita metro emitters. | Transport | 100 largest metropolitan areas of the US | Brown *et al.* (2008) |

| Parameter | Indicator | Relationship | Affecting sectors | Evidence | Source |
|---|---|---|---|---|---|
| *Settlement organization (explained in section 4.7)* | | | | | |
| Urban morphology | Spatial structures – mono-centric, polycentric, composite (or multiple-nuclei) models and urban village model | Urban spatial structures play a major role in determining not only population densities, but also the transportation mode (e.g. the relative importance of public vs. private modes) and with it cities' levels of energy use and GHG emissions. | Transport and residential and commercial | Four urban structures or forms can be distinguished – mono-centric, represented by such cities as New York (US), London (UK), Mumbai (India) and Singapore, polycentric, exemplified by such cities as Houston (US), Atlanta (US) and Rio de Janeiro (Brazil). Composite (or multiple-nuclei) and urban village models do not exist in the real world, but can be found only in urban master plans | UN-Habitat (2011, p. 56) and Bertaud *et al.* (2009) |
| Location of population and economic activities | Polycentricity of urbanization | Changes in the localization patterns of population and economic activities, as well as increased private automobile use, lie behind a transition in the urban form of the largest cities from a city-based to a region-based pattern. The polycentric pathway of urbanization is associated with carbon-relevant consequences, especially when it is not accompanied by public transportation policies. | Transport | Latin American cities (selected) – in the last two decades, Buenos Aires, Santiago and Mexico City have experienced a polycentric urban expansion of first- and second-order urban localities sprawling along major highways and functionally linked to the main city | Lankao (2007) |

*(continued)*

Table 4.2 (continued)

| Parameter | Indicator | Relationship | Affecting sectors | Evidence | Source |
|---|---|---|---|---|---|
| Land use (explained in section 4.8) | | | | | |
| Land use | Diversity and segregation potential | Mixed land use, diversity and greening, among others, are equally significant design concepts to density, in achieving sustainable urban form. A more complex relationship between land use and GHG emissions involves a model that also takes into account landscape impacts, deforestation, carbon sequestration by soils and plants, urban heat island, infrastructure impacts, transportation-related emissions, waste management-related emissions, electric transmission and distribution losses, and buildings (residential and commercial). There are complex relationships between these factors – for example, denser residential areas may have lower levels of car use, but simultaneously present fewer options for carbon sequestration. | LULUCF | Municipalities in New Jersey | Andrews (2008), Jabareen (2006, cited in UN-Habitat, 2011, p. 55) |
| Land use | Albedo and heat flux | Replacing natural vegetation with roads and buildings often decreases the surface albedo and alters the local surface energy balance, increasing sensible heat flux and decreasing latent heat flux. Although this effect has been verified with respect to local "urban heat islands" (UHI), the empirical evidence linking urban land use to regional or global climate change does not appear to be robust. | LULUCF | | Kueppers et al. (2008, p. 251, cited in Martine, 2009) |

| Parameter | Indicator | Relationship | Affecting sectors | Evidence | Source |
|---|---|---|---|---|---|
| Land use | Economic base | Variations in the proportion of greenhouse gas emissions that can be attributed to different sectors reflect the economic base of different cities (whether this is primarily industrial or service-oriented). | LULUCF | 12 global cities, including Barcelona, Glasgow, London, District of Columbia, New York, Toronto, Rio de Janeiro, São Paulo, Beijing, Seoul, Shanghai and Tokyo | Dodman (2009) |
| Land use under transport | Road area per capita and number of vehicles per km road length | Only temporal variations/trends of the indicator studied for sample cities from 1960 to 2000. | LULUCF | Four Asian megacities: Beijing, Shanghai, Tokyo and Seoul, no causal relationship established with GHG emissions | Dhakal (2004) |
| Predominant land use | Density and distribution of settlement | Variations in the proportion of greenhouse gas emissions that can be attributed to different sectors reflect the economic base of different cities (whether this is primarily industrial or service-oriented). | LULUCF | 12 global cities, including Barcelona, Glasgow, London, District of Columbia, New York, Toronto, Rio de Janeiro, São Paulo, Beijing, Seoul, Shanghai and Tokyo | Dodman (2009) |
| Predominant land use | Mixed land use | Mixed land use (residential and industrial, or residential and commercial, etc.) results in different energy use than does segregated land use. Urban zoning policies and industrial relocation from city centres to peri-urban areas in Asian cities significantly influence travel demand and energy use. Similarly, the energy use patterns in commercial cities are different from those in industrial cities. | Transport and energy use in residential, commercial and industrial sectors | Four Asian megacities: Beijing, Shanghai, Tokyo and Seoul | Dhakal (2004) |

sectors excluding electricity) corresponds quite closely to *heating degree days* (using an 18.0° C base temperature). This category excludes electricity used for heating. The linear fit has a statistically significant gradient (*t stat* 4.28 and an $R^2$ of 0.70). Denver and Toronto have the greatest consumption at 73.5 and 58.9 GJ/cap, respectively, while Cape Town and Barcelona each consumes less than 16 GJ/cap. The GHG emissions for heating and industrial fuel use largely follow the same pattern of energy consumption:

**Suggested indicator:** (X1) climate type

**Method of measurement:** Residential and commercial energy consumption (in M Kwh)

Geophysical conditions, such as the location of a settlement and its latitude and longitude, affect its climate, which in turn influence its energy requirements for heating/cooling and lighting. India predominantly hosts *tropic, sub-tropic* and *sub-temperate* climate, which is more towards the warmer side, hence *heating degree days* is not a commonly measured indicator by any agency. Similarly, data on lighting are also not known to be assimilated by any agency or institution.

From the perspective of planning, the climate conditions of a city/region could be deciphered either from the Koppean system of climate classification or more so from the climate zones identified by the National Building Code (NBC). Since this study intends to study GHG emissions from energy involved in the heating/ cooling and lighting of buildings, adherence to NBC climate zones is preferable. The code uses five major climate types, namely *hot dry, warm humid, composite, temperate* and *cold* (see Figure 4.2). But what if one wants to assess emissions numerically from the point of view of energy consumption? In that case, exclud- ing emissions from industrial or manufacturing sources, the city emissions are essentially from residential and commercial activities. Hence, *residential and commercial energy consumption* (in M KWh) could be a suitable measure to quantify variations in climate type on a numerical scale.

## 4.4 Regional connectivity

The location of a city within the region and its interconnections with other set- tlements therein could influence its carbon footprint. Urban areas with regional, national or international significance are likely to have greater interlinkages and activities that cause GHG emissions – for instance, cities that are either road/ railroad hubs, seaports and airports, or at intersection of highways, trade routes, shipping lanes or flight paths. Kennedy *et al.* (2009b) define this as *gateway status* of the city (see Table 4.2). In addition, the geographical location in rela- tion to natural resources influences the fuels that are used for energy generation, and hence the levels of GHG emissions. This is a factor of transportation costs: where a more efficient source of fuel is available in close proximity to the city or town, it can be used more economically. For example, urban areas that are able to

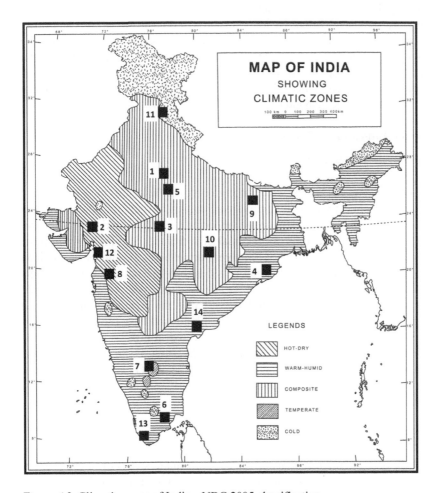

*Figure 4.2* Climatic zones of India – NBC 2005 classification

Source: NBC, Bureau of Indian Standards 2005 (www.bis.org.in/sf/nbc.htm)

Note: Sample city notations: (1) Agra; (2) Ahmedabad; (3) Bhopal; (4) Bhubaneswar; (5) Gwalior; (6) Madurai; (7) Mysore; (8); Nashik; (9) Patna; (10) Raipur; (11) Shimla; (12) Surat; (13) Thiruvananthapuram; (14) Vijayawada.

draw on nearby sources of natural gas will emit a smaller volume of GHGs for a given amount of energy than areas that rely on coal for energy (UN-Habitat, 2011, p. 52). The same logic may apply to non-fossil energy sources too.

The potential for using renewable sources of energy – and the reductions in GHG emissions associated with this – are also affected by locational factors. Some renewable energy is entirely reliant on natural resources – for example, the availability of large rivers for hydroelectric generation. Rio de Janeiro and São Paulo (in Brazil) have low levels of emissions from electricity generation for this reason (UN-Habitat, 2011, p. 53). In the case of India, unfortunately, the energy supplied to cities in various forms, particularly electricity, could not be definitively associated with the energy resource regions for its origins and flows. With the presence of a national and interconnected grid, it is virtually impossible to obtain empirical data on power supplied to numerous settlements from different thermal power, hydropower or DG stations at various points of time. Moreover, this parameter only influences the supply-side function and inadequately represents the consumptive demand of cities:

**Suggested Indicator:** (X2) number of linkages

**Method of measurement:**

The gateway status of a city within a region could be determined by its land, water and air connectivity. Hilly cities are acutely disadvantaged when compared to ones located on plains. Urban areas connected with a rail station or a network of railroads (junction) exhibit its regional significance. Similarly, water-based accessibility, through inland ports on rivers or seaports, promotes the regional connectivity of a settlement. Opening of skies to domestic and/or international flight carriers further enhances the city's gateway status. Accordingly, the greater the number of land, water and air linkage of a city, the greater the probability of its energy-induced carbon emissions. A scoring method (see Table 4.3) could be used to quantify regional linkages of a city located in hills or plains by simple addition of scores. On this basis, the capital city, New Delhi, which is located in plains (1), having a rail junction (2), access to a river (1) and international air linkages (2), would have a total score of 6. Similarly, Shimla and Kolkata would have a total score of 3 and 7, respectively.

## 4.5 Settlement scale

For the purpose of both research and policy, scale matters: "As different phenomena take place at different spatial scales, the preferred spatial scale depends on the analysis undertaken" (van Vuuren *et al.*, 2007, p. 114). Data for large regions or groups of countries are sufficient for global assessments and scenarios such as those developed in the *Global Environment Outlook* or the *Special Report on Emissions Scenarios* studies. Global models, however, hide variations between and within countries – for instance, energy intensity may vary considerably from one country to another (Dao and van Woerden, 2009). Besides 'real' physical

*Table 4.3* Scoring method to quantify a city's regional connectivity

| | | *Rail* | | *Navigation* | | *Air linkage* | |
|---|---|---|---|---|---|---|---|
| *Location* | *Access through land (road)* | *Station* | *Junction* | *Access to river/sea* | *Major port* | *Domestic* | *International* |
| Hills | 0 | 1 | 2 | 1 | 2 | 1 | 2 |
| Plains | 1 | 1 | 2 | 1 | 2 | 1 | 2 |

climate data, it is important to consolidate and improve authoritative data collections and compilations in the socio-economic and natural resources realms, using statistical surveys, as well as other sources such as satellite imagery – the end goal being to have proper, authoritative data in place to assess and address climate change issues adequately at all levels. In terms of population, the two key variables, of course, are size and distribution – both to estimate absolute figures and to derive per capita data. *This is particularly important for cities in developing countries, where large-size cities seem to have a disproportionately higher emission share compared to their population* (Dhakal, 2008).

Across cities, existing studies (see Table 4.2) point to a large variation in the scale of total and per capita emissions. Comparison of over 50 cities (Carney *et al.*, 2009; Chen *et al.*, 2009; Dhakal, 2004; Kennedy *et al.*, 2009a) points that such differences emerge from the nature of emission sources, urban economic structures (balance of manufacturing vs. service domination), local climate and geography, stage of economic development, fuel mix, state of public transport, and others (Dhakal, 2008). It is further argued that big cities also seem to evade the usual developing and developed country substantiation for the above reasons. Thus, per capita $CO_2$ emissions of certain cities in the developing world, such as Beijing, Shanghai, Tianjin and Bangkok, are higher than those of Tokyo, New York City and Greater London. As a result, it is often difficult to devise criteria and to compare the GHG emissions and GHG performance of cities globally.

Brown *et al.* (2008) provide an assessment of the carbon footprint of urban residents in the United States that takes into account highway transportation and energy consumption in residential buildings (excluding emissions from commercial buildings, industry or non-highway transportation). Their findings state that the average resident of metropolitan areas in the United States has a smaller carbon footprint (2.24 tons) than the average US citizen (2.6 tons), and that despite housing two-thirds of the US population and three-quarters of its economic activity, the 100 largest metropolitan areas in the US emitted just 56% of the country's carbon emissions from highway transportation and residential buildings in 2005. However, there is substantial variation between these metropolitan centres, with residents of Lexington, Kentucky (the highest-emitting metropolitan area), emitting 2.5 times more carbon than residents of Honolulu, Hawaii – with development patterns, the fuels used to generate electricity, weather and the availability of rail transit having an important effect on these variations. *The authors conclude that large metropolitan areas offer greater energy and carbon-efficiency than non-metropolitan areas* (Brown *et al.*, 2008, p. 7). This is often understood as "economies of scale" in urban parlance. It would be pertinent to see how GHGs from Indian cities respond as the city size changes:

**Suggested indicators:** (X3) urban area and (X4) urban population

**Method of measurement:**

From an areal point of view, India does not have a scale-based definition or classification of settlements. Area figures are available for differently defined

jurisdictional boundaries/areas of the city, such as 'planning area', administered by the development authority or city-region authority, or 'municipal area',
administered by a local municipal corporation, council, board, etc. It is observed
that planning areas generally encompass huge tracts of rural areas, population,
economic activities and land use, and as such do not truly represent urban characteristics. As such, the use of 'municipal area' to measure the urban area is
preferred. Meanwhile, the Census of India has a defined urban hierarchy of settlements based on population criteria (Class I–VI and urban agglomeration). In
addition, to maintain consistency between the area and its census population, the
administrative unit (i.e. municipal corporation/council/ board, etc.) has been taken
into consideration. Since both 'area' and 'population' are numerical variables, it is
prudent to measure and plot them on a numerical scale.

## 4.6 Urban form – compact vs. sprawl

Urban form and density are associated with a range of social and environmental consequences. On the one hand, the extremely high densities of many cities
in developing countries – particularly in informal settlements and other slums –
result in increased health risks, and high levels of vulnerability to climate change
and extreme events. On the other hand, the extremely low densities of many
suburban areas in North America are associated with high levels of household
energy consumption as a result of sprawling buildings and extensive car usage
(UN-Habitat, 2011, p. 54).

At the global level, there is strong evidence that, in general, urban densities
have been declining over the past two centuries (United Nations Population Fund,
2007). Perhaps the most detailed and compelling assessment of this phenomenon
is provided by a World Bank report that records the decline in the average density
of developed country cities from 3,545 to 2,835 people/sq km between 1990 and
2000. During the same period, the average urban population density in developing countries declined from 9,560 to 8,050 people/sq km (Angel *et al.*, 2005). The
reduction in urban densities is likely to continue into the future. It is estimated that
the total population of cities in developing countries will double between 2000
and 2030, but their built-up areas will triple from approximately 200,000 sq km
to approximately 600,000 sq km. During the same period, the population of cities
in developed countries is projected to increase by approximately 20%, while their
built-up areas will increase by 2.5 times, from approximately 200,000 sq km to
approximately 500,000 sq km.

Evidence suggests that there are several causation, indicators and affecting
sectors by virtue of urban form (see Table 4.2). Spatially compact and mixed use
urban developments have several benefits in terms of GHG emissions, which
includes reduced costs for heating and cooling resulting from smaller homes
and shared walls in multi-unit dwellings, lesser line losses related to electricity transmission and distribution, along with the possibility of using micro-grids
at the local level, reduced average daily vehicle kilometres travelled in freight
deliveries and by private motor vehicles per capita. Population density increases

accessibility to such destinations as stores, employment centres and theatres (Newman and Kenworthy, 1999; UN-Habitat, 2011, p. 54).

The high concentrations of people and economic activities in urban areas can lead to economies of scale, proximity and agglomeration – all of which can have a positive impact upon energy use and associated emissions, while the proximity of homes and businesses can encourage walking, cycling and the use of mass transport in place of private motor vehicles (Satterthwaite, 1999). Some researchers suggest that each doubling of average neighbourhood density is associated with a decrease of 20–40% in per household vehicle use, with a corresponding decline in emissions (Gottdiener and Budd, 2005). An influential paper published in 1989 suggested that gasoline use per capita declines with urban density (Newman and Kenworthy, 1989), although this relationship weakens once GDP per capita is brought into consideration (World Bank, 2010), as shown in Figure 4.3.

A recent study of GHG emissions in Toronto (Canada) deals with the issue of density explicitly. The study depicts both the overall patterns of GHG emissions and examines how these vary spatially throughout the Toronto Census Metropolitan Area: as the distance from the central core increases, private motor vehicle emissions begin to dominate the total emissions (VandeWeghe and Kennedy, 2007). This pattern is supported by an earlier study, which found that low-density suburban development in Toronto is 2 to 2.5 times more energy- and GHG-intensive than high-density urban core development on a per capita basis (Norman *et al.*, 2006). Density may also affect household energy consumption. More compact housing uses less energy for heating. For example, households in the US living in single-family detached housing consume 35% more energy for heating and 21% more for cooling than comparable households in other forms of housing (UN-Habitat, 2011, p. 55).

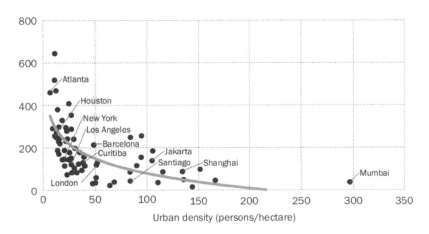

*Figure 4.3* City densities and their transport emissions in kg per capita

Source: World Bank (2010)

Dense urban settlements can therefore be seen to enable lifestyles that reduce per capita GHG emissions through concentration of services that reduce the need to travel large distances, better provision of public transportation networks, and constraints on the size of residential dwellings imposed by the scarcity and high cost of land. Yet, conscious strategies to increase urban density may or may not have a positive influence on GHG emissions and other environmental impacts. Many of the world's most densely populated cities in South, Central and South East Asia suffer severely from overcrowding, and reducing urban density will meet a great many broader social, environmental and developmental needs. In addition, a variety of vulnerabilities to climate change are also exacerbated by density. Coastal location, exposure to the urban heat island effect, high levels of outdoor and indoor air pollution, and poor sanitation are associated with areas of high population density in developing country cities (Campbell-Lendrum and Corvalan, 2007). However, these also provide clear opportunities for simultaneously improving health and cutting GHG emissions through policies related to transport systems, urban planning, building regulations and household energy supply (UN-Habitat, 2011, p. 55).

Urban density influences the amount of energy used in private passenger transport, and therefore also has a significant effect on greenhouse gas emissions. They allow many more journeys to be made by walking or bicycling, greater use of public transport and a high-quality service more feasible. Many prosperous European cities, with among the world's highest quality of life, have one-fifth of the gasoline use per person of the USA's less compact, more car-dependent cities (Newman, 2006). Most European cities have high-density centres where walking and bicycling are preferred by much of the population, especially where good provision is made for pedestrians and bicyclists (including public transport that can accommodate bicycles). Dodman (2009) argues that, with the exceptions of the Chinese cities (having manufacturing functions) included rankings of high emissions, the most densely populated cities utilize less energy for private passenger transport and generally have lower GHG emissions per capita.

**Suggested indicator:** (X5) Urban density

**Method of measurement:**

Though urban morphology or form could be analysed through several indicators such as building heights, built open spaces, etc., density is by far the most frequent and reliable indicator used to measure intensity of built form and its variations. In the case of India, there is no standard set of definition or regulation to classify urban densities (such as low-density, medium-density or high-density cities or localities). Hence, density is measured on a continuous numerical scale, using the following formula:

$$\text{urban density} = \frac{\text{census population of the municipal area}}{\text{municipal area (in sq km)}}$$

As in the case of quantifying the USP – *settlement scale*, data on urban area and urban population is considered for the relevant municipal unit/boundary. Accordingly, the *density* would also be computed for the corresponding municipal area.

## 4.7 Spatial organization/settlement structure

Urban spatial structures play a major role in determining not only distribution of population and work areas, but also the transportation mode (e.g. the relative importance of public vs. private modes), and with it cities' levels of energy use and GHG emissions (see Table 4.2 to review causation, indicators, affecting sectors and evidence of this parameter). While urban structures do evolve with time, driven by changes in the localization of economic activities, real estate developments and population, their evolution is slow and can seldom be shaped by design. The larger the city, the less it is amenable to change its urban structure (UN-Habitat, 2011, p. 56). Hence, in one sense of the word, spatial structure could be understood as the location of major economic activities with respect to residential districts within the city.

According to Bertaud *et al.* (2009), four urban structures or forms can be distinguished. In the first, *mono-centric*, represented by such cities as New York, London, Mumbai and Singapore, most economic activities, jobs and amenities are concentrated in the central business district (CBD). Here, authorities should focus on promoting public transport as the most convenient transport mode, as most commuters travel from the suburbs to the CBD. In the second, *polycentric*, exemplified by such cities as Houston, Atlanta and Rio de Janeiro, few jobs and amenities are located in the centre, and most trips are from suburb to suburb. A very large number of possible travel routes exists, but with few passengers per route. Therefore, public transport is difficult and expensive to operate, and individual means of transportation or collective taxis are, and should be, promoted as the more convenient transportation options for users.

The third form, *composite* (or *multiple-nuclei*) model, is the most common type of urban spatial structure, containing a dominant centre together with a large number of jobs located in the suburbs. Most trips from the suburbs to the CBD are made and should be promoted by public transport, while trips from suburb to suburb are made with individual cars, motorcycles, collective taxis or minibuses. The fourth, also called the *urban village* model, is not believed to exist in real-world cities, but in urban master plan proposals. In this model, urban areas contain many business centres, commuters travel only to the centre that is the closest to their residence, and have more opportunities to walk or bicycle to work. It is an ideal case because it requires less transportation and roads, thus, in theory, dramatically reducing distances travelled, energy use and, as a consequence, GHG emissions and other air pollutants. However, it is argued to be feasible, as "it implies a systematic fragmentation of labour markets which would be economically unsustainable in the real world", or, simply put, diseconomies of scale. In the last two decades, Latin American cities such as

Buenos Aires, Santiago and Mexico City have experienced a polycentric urban expansion of first- and second-order urban localities sprawling along major highways and functionally linked to the main city. The polycentric pathway of urbanization is associated with carbon-relevant consequences, especially when it is not accompanied by public transportation policies (Lankao, 2007; Pablo, 1999). In addition to the above recognizable structures, unplanned or organic models also exist in practice.

In the second sense of the word, spatial organization could be understood from the city network, the most fundamental being linear, grid-iron, radial, concentric, radio-concentric and non-structured without any recognizable pattern. Radial cities are the most simple and common ones, usually being the origin of most cities with a central hub, and hence could be co-related with the mono-centric model. Linear cities most often evolve from the central node (mono-centric model) as cities grow under topographic constrains of prominent physical features such as mountain chains, hill outcrops, sea coasts or river shores. Grid-iron and concentric cities exist as conceptual models mostly in isolated town-planning schemes, but hardly exist at full-grown urban scale. For instance, grid-irons are hardly distinguishable from linear models, while concentric networks are interspersed with necessary radials evident in radial cities. In practice, radio-concentric networks are complex structures formed as cities evolve from simple radial, concentric or linear networks. They are the most common and frequently found spatial structures and could be related to polycentric and multi-nuclei models. Meanwhile, non-structured networks are associated with organic and unplanned forms of urban development.

**Suggested indicators:** (X6) City structure and (X7) Number of arteries

**Method of measurement:**

The spatial organization of Indian cities could be analysed by a combined study of both their city structure (linear, radial, radio-concentric, organic) and its complexity (measured by the number of major arteries or radials) through the following inventory (Table 4.4). For quantitative assessment of energy use invested in such a spatial structure, annual transportation fuel consumption (KL) is used as a substitute indicator.

*Table 4.4* Method to inventorize a city's spatial organization/structure

| **Indicator 1: city structure** | *Linear (L)* | *Radial (R)* | *Radio-concentric (RC)* | *Organic (O)* |
|---|---|---|---|---|
| **Indicator 2: number of major arteries** | | 0, 1, 2, 3, 4, 5, 6, 7 . . . $n$ | | |
| **Sample notation** | L3: Linear structure with three main arteries | | | |
| | RC5: Radio-concentric with five main arteries | | | |

## 4.8 Land use

Land use changes are considered a first-order climate-forcing factor: around 31% of all greenhouse gas emissions are reputed to arise from the land use sector (Scherr and Sthapit, 2009, p. 32). Research compiled by UN-Habitat suggests that mixed land use, diversity and greening, among others, are equally significant design concepts to density, in achieving sustainable urban form (Jabareen, 2006; UN-Habitat, 2011). A more complex relationship between land use and GHG emissions involves a model that also takes into account landscape impacts (deforestation, carbon sequestration by soils and plants, urban heat island), infrastructure impacts, transportation-related emissions, waste management-related emissions, electric transmission and distribution losses, and buildings (residential and commercial). There are complex relationships between these factors – for example, denser residential areas may have lower levels of car use, but simultaneously present fewer options for carbon sequestration (Andrews, 2008). Definite research demonstrating the relationship between land use and GHG emissions is cited in Table 4.2.

In terms of functional use of land, extractive activities (such as mining and lumbering) and energy-intensive manufacturing are obviously associated with higher levels of emissions – especially when the energy for these is supplied from fossil fuels (UN-Habitat, 2011, p. 57). However, there are fewer of these activities in many cities in developed countries, as lower transportation costs and the lower cost of labour elsewhere have encouraged industries to relocate. In London, for example, industrial emissions halved between 1990 and 2006, as industrial activity has relocated to other parts of the UK or overseas (Mayor of London, 2007). The influence of the urban economy on patterns of emissions can be seen in the large variations in the proportion of a city's GHG emissions that can be attributed to the industrial sector. Industrial activities in many rapidly industrializing developing countries, such as China, are responsible for a large proportion of urban GHG emissions. Indeed, while 12% of Chinese emissions were due to the production of exports in 1987, this figure had increased to 21% in 2002 and 33% (equivalent to 6% of total global $CO_2$ emissions) in 2005 (Weber *et al.*, 2008). As such, study of industrial land use becomes paramount to trace GHG emissions.

In contrast, GHG emissions from the industrial sector in cities elsewhere are much lower, generally reflecting a transition to service-based urban economies. Industrial activities account for just 0.04% in Washington, DC (largely because of the narrow spatial definition of the District of Columbia), 7% in London, 9.7% in São Paulo, Brazil, and 10% in Tokyo and New York (compared to 29% for the US as a whole). The declining importance of industry in causing urban emissions is evident in several cities. In Rio de Janeiro, the industrial sector's proportion of emissions declined from 12% in 1990 to 6.2% in 1998, and in Tokyo it declined from 30% to 10% during the last three decades (compiled from Newman, 2006).

Although the changes in land use brought about by urban growth are routinely cited as a major factor in the growth of GHG emissions, the actual degree of this impact appears open to questions. In principle, "replacing natural vegetation with roads and buildings often decreases the surface albedo and alters the local surface energy balance, increasing sensible heat flux and decreasing latent heat flux" (Kueppers *et al.*, 2008, p. 251). Although this effect has been verified with respect to local "urban heat islands" (UHI), the empirical evidence linking urban land use to regional or global climate change does not appear to be robust (Martine, 2009, p. 18). As such, there is not much empirical research or evidence on the carbon sequestrization potential of different land use in diverse locations and biomes/vegetative regions of the world. According to a Malaysian compendium, *Low Carbon Cities Framework Assessment System* (Kementerian Tenaga, 2011), one acre of development in infill and brownfield area emits 7,000 kg $CO_2$ emission – savings of 3,000 kg of $CO_2$ compared to greenfield development. Similarly, areas under buildings, roads or paved surfaces in cities could impact GHG emissions. Meanwhile, 1 hectare with 0.1 m thickness of asphalt emits 70,150 kg of $CO_2$/year, whereas 1 hectare with 0.1m thickness of concrete pavement emits 15,800 kg of $CO_2$/year. Box 4.1 displays the carbon sequestrization potential of some common land use/land covers.

---

**Box 4.1    Carbon sequesterization potential of different land use/land covers**

- A tropical forest absorbs 5.5 kg of $CO_2$/year.
- 1 hectare of tropical forest captures 4.3 t $CO_2$/year.
- 1 hectare of tropical wetland absorbs 1.48 t $CO_2$/year.
- 1 tree absorbs approximately 1,000 kg of $CO_2$.
- 1 acre of trees stores 2,600 kg of carbon/year, where tree cover for urban area is about 204 trees/acre; for forests, it is about 480 trees/acre.

Source: www.coloradotrees.org and www.conservationfund.org/gozero

---

Accordingly, the three most crucial indicators to study within land use are: (1) industry, which causes industrial emissions from stationary sources; (2) transportation/ roads, which is a prime contributor to vehicular emissions; and; (3) recreational, greens, parks, etc., which are responsible for sequesterization of emissions in the city.

**Suggested indicators:** (X8) Industrial land use, (X9) Transportation land use and (X10) Recreational land use

**Method of measurement:**

The land use under industrial, transportation and recreational functions could be analysed by evaluating their area under actual or relative use in the city, through the following inventory (see Table 4.5).

*Table 4.5* Method to quantify a city's industrial, transportation and recreational land use

| Indicator | Method of measurement | Source |
| --- | --- | --- |
| **Industrial land use** | Area under industrial land use at the city level, expressed in sq km or percentage | Land use plan, master plan, city development plan, etc. of the relevant city |
| **Transportation land use** | Area under roads, railway land, etc. at the city level, expressed in sq km or percentage | Land use plan, master plan, city development plan, comprehensive mobility plan, comprehensive traffic and transportation plan, etc. of the relevant city |
| **Recreational land use** | Area under recreation, parks, green, urban forestry, ridge areas, conserved, protected areas, district parks, etc. at the city level, expressed in sq km or percentage | Land use plan, master plan, city development plan, etc. of the relevant city |

## Basic questions

1 Using IPAT + spatial framework, enlist/discuss the main causations and drivers of GHGs from a particular sector (say, transport or energy).
2 What are the spatial causations of GHGs in the residential and commercial sectors?
3 Based on the literature, list six main urban spatial parameters that influence GHGs. Explain one in depth to demonstrate how it contributes.
4 Discuss three main subcategories of land use in a city that can possibly influence its GHG emissions.

## Advanced questions

1 Deliberate upon how USP–regional connectivity (i.e. gateway status of a city) is related to theories in regional economics/planning (such as Walter Christaller's central place theory).
2 What is the impact of agglomeration effect on a city's carbon footprint. All else being equal, does a city's total population or its density have greater control over its GHG emissions? Can this be related to *Jevons Paradox*?
3 Substantiate with evidence that compact settlements are effective in reducing a city's carbon footprint. How is this phenomenon relevant to theories on urban form?
4 Appraise the USP (of GHG emissions) – spatial organization/settlement structure with respect to classic urban theories on location of economic activities, transportation and city structure (Burgess' concentric zone model, Homer Hoyt's sector model, Harris and Ullman's multiple nuclei model, etc.).

**Do it yourself exercises**

1   Taking your own household as a microcosmic city, and consumption of energy and material resources as a substitute indicator of your family's GHG emissions, start listing your daily activities, such as eating, working, travelling, shopping, recreation, etc., and generation of waste. Reason out which of these activities or consumption of energies are driven by the *size* of your family (though some may actually be negatively correlated due to sharing of items, space, etc.), the *affluence level* and *efficiency level* of resource consumption or waste generation. Other than the above, list down those factors that, by virtue of space and spatial location/disposition, could inflate or depreciate the impacts (for instance, the area or total volume of your house, distance to the office, or groceries, etc.).

**Suggested reading**

*Normative spatial causations of GHG emissions*

1   Bertaud, A., Lefevre, B. and Yuen, B. (2009). GHG emissions, urban mobility and efficiency of urban morphology: A hypothesis. Paper prepared for the 5th Urban Research Symposium, *Cities and climate change: Responding to an urgent agenda*, 28–30 June, Marseille, France.
2   Dhakal, S. (2004). *Urban energy use and greenhouse gas emissions in Asian megacities*. Kitakyushu, Japan: Institute for Global Environmental Strategies.
3   Dhakal, S. (2010). GHG emissions from urbanization and opportunities for urban carbon mitigation. *Current Opinion in Environmental Sustainability, 2*, 277–83.
4   Gottdiener, M. and Budd, L. (2005). *Key concepts in urban studies.* London: Sage.
5   Kennedy, C., Steinberger, J., Gasson, B., Hansen, Y., Hillman, T., Havranek, M., Paraki, D., Phdungsilp, A., Ramaswami, A. and Mendez, G.V. (2009b). Greenhouse gas emissions from global cities. *Environmental Science and Technology, 43*, 7297–302.
6   Newman, P. and Kenworthy, J. (1989). Gasoline consumption and cities: A comparison of US cities with a global survey. *Journal of the American Planning Association, 55*(1), 24–37.
7   UN-Habitat (2011). *Cities and climate change: Global report on human settlements.* London/Washington, DC: Earthscan & UNCHS.
8   World Bank (2010). *Cities and climate change: An urgent agenda.* Washington, DC: IRDC.

**References**

Andrews, C. (2008). Greenhouse gas emissions along the rural–urban gradient. *Journal of Environmental Planning and Management, 51*(6), 847–70.
Angel, S., Sheppard S.C. and Civco, D.L. (2005). *The dynamics of global urban expansion.* Washington, DC: World Bank.
Baldasano, J.M., Soriano, C. and Boada, L. (1999). Emission inventory for greenhouse gases in the City of Barcelona, 1987–1996. *Atmospheric Environment, 33* (23), 3765–3775.
Barker, T., Bashmakov, I., Bernstein, L., Bogner, J., Bosch, P., Dave, R. and Halsnaes, K. (2007). Climate Change 2007: Mitigation. *Metz, B,* 619–690.
Bertaud, A., Lefevre, B. and Yuen, B. (2009). GHG emissions, urban mobility and efficiency of urban morphology: A hypothesis. Paper prepared for the 5th Urban Research

Symposium, *Cities and climate change*: *Responding to an urgent agenda*, 28–30 June, Marseille, France.

Brown, M., Southworth, F. and Sarzynski, A. (2008). *Shrinking the carbon footprint of metropolitan America.* Washington, DC: Brookings Institution. Retrieved from www. brookings. edu/papers/2008/05_carbon_ footprint_sarzynski.aspx on 2 December 2008.

Campbell-Lendrum, D. and Corvalan, C. (2007). Climate change and developing country cities: Implications for environmental health and equity. *Journal of Urban Health, 84*(1), 109–17.

Carney, S., Green, N., Wood, R. and Read, R. (2009). Greenhouse gas emissions inventories for eighteen European regions, EU CO2 80/50 Project Stage 1: Inventory formation. *The greenhouse gas regional inventory protocol (GRIP).* Manchester: Centre for Urban and Regional Ecology, School of Environment and Development, University of Manchester.

Chen, K.S., Li, H.C., Wang, H.K., Wang, W.C. and Lai, C.H. (2009). Measurement and receptor modeling of atmospheric polycyclic aromatic hydrocarbons in urban Kaohsiung, Taiwan. *Journal of Hazardous Materials, 166*, 873–9.

Dao, H. and van Woerden, J. (2009). Population data for climate change analysis. In J.M. Guzman, G. Martine, G. McGranaghan, D. Schensul and C. Tacoli (Eds.), *Population dynamics and climate change* (pp. 218–38). New York/London: UNFPA & IIED.

Dhakal, S. (2004). *Urban energy use and greenhouse gas emissions in Asian megacities.* Kitakyushu, Japan: Institute for Global Environmental Strategies.

Dhakal, S. (2008) Climate change and cities: The making of a climate friendly future. In P. Droege (Ed.), *Urban energy transition: From fossil fuels to renewable power* (pp. 173–92). Oxford: Elsevier Science.

Dhakal, S. (2010). GHG emissions from urbanization and opportunities for urban carbon mitigation. *Current Opinion in Environmental Sustainability, 2*, 277–83.

Dodman, D. (2009). Blaming cities for climate change? An analysis of urban greenhouse gas emissions inventories, *Environment and Urbanization, 21*(185). doi:10.1177/ 0956247809103016.

Glaeser, E. and Kahn M. (2008). *The greenness of cities: Carbon dioxide emissions and urban development.* Harvard Kennedy School: Taubman Center for State and Local Government – Working Paper.

Gottdiener, M. and Budd, L. (2005). *Key concepts in urban studies.* London: Sage.

Jabareen, Y. (2006). Sustainable urban forms: Their typologies, models, and concepts. *Journal of Planning Education and Research, 26*(1), 38–52.

Kementerian Tenaga (2011). *Low carbon cities framework and assessment system.* Kuala Lumpur: Teknologi Hijau dan Air-KeTTHA, Government of Malaysia.

Kennedy, C.A., Ramaswami, A., Carney, S., Dhakal, S. (2009a). Greenhouse gas emission baselines for global cities and metropolitan regions. Paper presented in the 5th Urban Research Symposium, *Cities and climate change*: *Responding to an urgent agenda*, 28–30 June, Marseille, France.

Kennedy, C., Steinberger, J., Gasson, B., Hansen, Y., Hillman, T., Havranek, M., Paraki, D., Phdungsilp, A., Ramaswami, A. and Mendez, G.V. (2009b). Greenhouse gas emissions from global cities. *Environmental Science and Technology, 43*, 7297–302.

Kueppers, L.M. *et al.* (2008). Seasonal temperature responses to land-use change in the western United States. *Global and Planetary Change, 60*(3–4), 250–64. Retrieved from http://faculty.ucmerced.edu/lkueppers/pdf/Kueppers%20et%20al%202008%20 land%20use%20&%20climate%20W%20US.pdf on 7 February 2014.

Martine, G. (2009). Population dynamics and policies in the context of global climate change. In J.M. Guzman, G. Martine, G. McGranaghan, D. Schensul and C. Tacoli (Eds.), *Population dynamics and climate change* (pp. 9–30). New York/London: UNFPA & IIED.

Mayor of London (2007). *Action today to protect tomorrow: The mayor's climate change action plan*. London: Greater London Authority.

Newman, P. (2006). The environmental impact of cities. *Environment and Urbanization, 18*(2), 275–95.

Newman, P. and Kenworthy, J. (1989). Gasoline consumption and cities: A comparison of US cities with a global survey. *Journal of the American Planning Association, 55*(1), 24–37.

Newman, P. and Kenworthy, J. (1999). *Sustainability and cities: Overcoming automobile dependence*. Washington, DC: Island Press.

Norman, J., MacLean, H.L. and Kennedy, C.A. (2006). Comparing high and low residential density: Lifecycle analysis of energy use and greenhouse gas emissions. *Journal of Urban Planning and Development, 132*(1), 10–21.

Pablo, C. (1999). Globalización y dualización en la Región Metropolitana de Buenos Aires. Grandes inversiones y restructuración socioterritorial en los años noventa, *EURE, 25*(76), 5–27.

Lankao, P.R. (2007). Are we missing the point? Particulars of urbanization, sustainability and carbon emissions in Latin American cities. *Environment and Urbanization, 19*, 157–75. doi:10.1177/0956247807076915.

Satterthwaite, D. (1999). The key issues and the works included. In D. Satterthwaite (Ed.), *The earthscan reader in sustainable cities* (pp. 3–21). London: Earthscan.

Satterthwaite, D. (2011). How urban societies can adapt to resource shortage and climate change. *The Royal Society A*, 369, 1762–1783.

Scherr, S.J. and Sthapit, S. (2009). Farming and land use to cool the planet. *2009 state of the world: Into a warming planet*. Washington, DC: Worldwatch. Retrieved from www.worldwatch.org/fi les/pdf/SOW09_chap3.pdf on 3 May 2014.

Sugar, L., Kennedy, C., and Leman, E. (2012). Greenhouse gas emissions from Chinese cities. *Journal of Industrial Ecology, 16* (4), 552–563.

UN-Habitat (2011). *Cities and climate change: Global report on human settlements*. London/Washington, DC: Earthscan & UNCHS.

United Nations Population Fund (2007). *State of the world population 2007: Unleashing the potential of urban growth*. New York: UNFPA.

van Vuuren, D.P., Lucas, P. and Hilderink, H. (2007). Downscaling drivers of global environmental change: Enabling use of global SRES scenarios at national and grid levels. *Global Environmental Change, 17*, 114–30. doi:http://dx.doi.org/10.1016/j.gloenvcha.2006.04.004.

VandeWeghe, J.R. and Kennedy, C. (2007). A spatial analysis of residential greenhouse gas emissions in the Toronto Census metropolitan area. *Journal of Industrial Ecology, 11*(2), 133–44.

Weber, C.L., Peters, G.P., Guan, D. and Hubacek, K. (2008). The contribution of Chinese exports to climate change. *Energy Policy, 36*(9), 3572–7.

World Bank (2010). *Cities and climate change: An urgent agenda*. Washington, DC: IRDC.

# 5   Spatial metrics and modelling

*If you can't measure it, you can't manage it.*

As elaborated in methodology (sections 1.9.2 and 1.9.3), the research conducts a pilot study on sample cities to test the strength of coefficient of correlation, its positive/negative sign and nature of equation (linear, polynomial, logarithmic and exponential) and suggests the appropriate unit of reporting. Based on this, a comprehensive OLS regression analysis is executed over 41 cities to arrive at a conclusive mathematical function/metric between urban spatial parameters and GHGs.

## 5.1 Sampling

The city selection in the pilot stage was based on stratified samples across diverse criteria that included regional climate, functional diversity (in terms of economic activity), urban hierarchy, variation in urban form, and spatial structure, as discussed below.

### 5.1.1 Regional climate

India has a large geographical diversity. Cities from different physio-geographical regions (i.e Great Himalayas, Gangetic Plains, Deccan Plateau, Indian Desert, Coastal Plains, Western Ghats, Central Highlands, etc.) need to be selected. As evident, local climate significantly influences the annual energy demand for heating and lighting in buildings. The National Building Code 2005 identifies five main climate zones in India, viz. hot dry, warm humid, composite, temperate and cold (illustrated in Figure 4.2). It is imperative to have samples from each climatic zone. In fact, urban settlements from all climatic zones have been selected effectively, in proportion to the expanse of each zone in India.

### 5.1.2 Functional diversity in terms of economic activities

Settlements exemplify differentiated functional or economic bases such as the primary, secondary or tertiary sector. Settlement types could accordingly be monofunctional (based either on agriculture, administrative, industrial, business/ finance, tourism, etc.) or multifunctional (a combination of these). In order to maintain consistency in data fields and comparability between them, samples are taken from multifunctional settlements. Highly specialized or industrialized towns dependent upon monofunction, such as Jamshedpur, Ranchi, Vishakhapatnam, etc., which overemphasize single-source emissions and tend to underrepresent emissions from remaining urban activities, are best avoided.

### 5.1.3 Urban hierarchy

Due diligence is given to include towns of different gateways, size/scale in terms of urban area, ranging from 19.55 to 466 sq km, and their population varying from 140,000 to 5,500,000, for Shimla and Ahmedabad, respectively. Megacities or higher-order settlements, such as New Delhi, Mumbai, Kolkata, Bangalore and Hyderabad, which tend to create severe data multiplicity, distortion and misrepresentation owing to complexities in size, function, jurisdictions, informal economies, etc., are purposely avoided. Though for the sake of better comprehension and comparisons, they are shown as reference points in selected analysis, as and where data permit.

### 5.1.4 Variation in urban form and spatial structure

As evident from discussions in the previous chapter, particularly sections 4.6 and 4.7, urban settlements exist in various forms and structural configurations, typically as linear, organic, radial, radio-concentric shapes with mono-centric or multi-nuclear structures and densities. Sample cities have been selected keeping in view that all spatial typologies are duly represented, as urban density ranges from 2,832 (Gwalior) to 12,519 (Patna).

Based on the above criteria, sample cities arranged according to different climate regions in India are (see also Figure 4.2 for the geographic location on the climate map of India):

**Hot dry:** Ahmedabad, Nashik, Surat

**Warm humid:** Bhubaneswar, Madurai, Thiruvananthapuram, Vijayawada

**Composite:** Agra, Bhopal, Gwalior, Patna, Raipur

**Temperate:** Mysore

**Cold:** Shimla

A data inventory of these 14 sample cities is prepared, which includes the following:

(a) City map from Google Maps.
(b) City population data from Census of India, 2001 and 2011.
(c) Spatial data: area, density, climate zone, settlement type, regional connectivity/gateway status, spatial structure, land use – industrial, transportation, parks and green areas.
(d) Emissions, energy and waste data – total emissions 2007, per capita emissions, electricity – residential, commercial and industrial, annual fuel consumption and municpal solid waste (MSW).

### 5.2 Data inventory

A data inventory for sampled cities is given below in their alphabetical order.

### 5.2.1 Agra

Population – 1,270,000 (2001), 1,574,542 (2011)

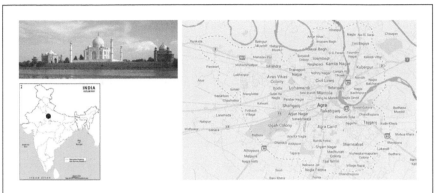

| Field | Value | Unit | Source |
|---|---|---|---|
| *Spatial data* | | | |
| Area | 188.40 | sq km | City Development Plan Agra |
| Density | 6,741.00 | population/sq km | |
| Climate zone | Composite | – | National Building Code 2005 |
| Settlement type | Multifunctional | – | |
| Regional connectivity/ gateway status | 5 | – | Indian Railways, Airport Authority of India, Port Trusts of India, Inland Water Authority of India |
| Spatial structure | Radial 8 | – | Google Earth 2015 |
| Land use <br> • Industrial <br> • Transportation <br> • Parks and green areas | <br> 00.62 <br> 25.20 <br> 00.42 | <br> % of total urban area | City Development Plan Agra |
| *Emissions, energy and waste data* | | | |
| Total emissions 2007 | 0.14 | Mt CO$_2$e | ICLEI 2009 |
| Per capita emissions | 0.66 | t CO$_2$e/year | ICLEI 2009 |
| Electricity <br> • Residential <br> • Commercial <br> • Industrial | <br> 53.92 <br> 25.30 <br> 153.00 | <br> M KWh | ICLEI 2009 |
| Annual fuel consumption | 17,727.00 | Kilolitres (KL) | ICLEI 2009 |
| MSW | 93.00 | Tons per day (tpd) | ICLEI 2009 |

*City image*: David Castor/Wikimedia Commons/public domain

*Map*: Map data @2015 Google

### 5.2.2 Ahmedabad

Population – 5,500,000 (2001), 6,352,254 (2011)

| Field | Value | Unit | Source |
|---|---|---|---|
| *Spatial data* | | | |
| Area | 466.00 | sq km | Revised Draft Development Plan of AUDA – 2011AD Part I, Vol 2 |
| Density | 11,803.00 | population/sq km | |
| Climate zone | Hot dry | – | National Building Code 2005 |
| Settlement type | Multifunctional | – | |
| Regional connectivity/ gateway status | 7 | | Indian Railways, Airport Authority of India, Port Trusts of India, Inland Water Authority of India |
| Spatial structure | Radio-concentric 16 | – | Google Earth 2015 |
| Land use<br>• Industrial<br>• Transportation<br>• Parks and green areas | <br>17.34<br>7.47<br>4.46 | % of total urban area | Revised Draft Development Plan of AUDA – 2011AD Part I, Vol 2 |
| *Emissions, energy and waste data* | | | |
| Total emissions 2007 | 6.78 | Mt $CO_2$e | ICLEI 2009 |
| Per capita emissions | 1.63 | t $CO_2$e/year | ICLEI 2009 |
| Electricity<br>• Residential<br>• Commercial<br>• Industrial | <br>1,334.22<br>948.12<br>2,266.62 | M KWh | ICLEI 2009 |
| Annual fuel consumption | 472,984.00 | Kilolitres (KL) | ICLEI 2009 |
| MSW | 2,242.00 | Tons per day (tpd) | ICLEI 2009 |

*City image*: Amcanada/Creative Commons Attribution 3.0 Unported License

*Map*: Map data @2015 Google

### 5.2.3 Bhopal

Population – 1,430,000 (2001), 1,795,648 (2011)

| Field | Value | Unit | Source |
|---|---|---|---|
| *Spatial data* | | | |
| Area | 284.00 | sq km | City Development Plan Bhopal |
| Density | 5,035.00 | population/sq km | |
| Climate zone | Composite | – | National Building Code 2005 |
| Settlement type | Multifunctional | – | |
| Regional connectivity/ gateway status | 5 | – | Indian Railways, Airport Authority of India, Port Trusts of India, Inland Water Authority of India |
| Spatial structure | Radio-concentric 7 | – | Google Earth 2015 |
| Land use <br> • Industrial <br> • Transportation <br> • Parks and green areas | <br> 9.00 <br> 15.00 <br> 13.00 | % of total urban area | City Development Plan Bhopal |
| *Emissions, energy and waste data* | | | |
| Total emissions 2007 | 0.74 | Mt $CO_2$e | ICLEI 2009 |
| Per capita emissions | 0.71 | t $CO_2$e/year | ICLEI 2009 |
| Electricity <br> • Residential <br> • Commercial <br> • Industrial | <br> 340.10 <br> 112.90 <br> 29.50 | M KWh | ICLEI 2009 |
| Annual fuel consumption | 108,212.00 | Kilolitres (KL) | ICLEI 2009 |
| MSW | 550.00 | Tons per day (tpd) | ICLEI 2009 |

*City image:* Shivamdwivedi82/Wikipedia/GNU Free Documentation License, Version 1.2

*Map*: Map data @2015 Google

### 5.2.4 Bhubaneswar

Population – 640,000 (2001), 840,683 (2011)

| Field | Value | Unit | Source |
|-------|-------|------|--------|
| *Spatial data* | | | |
| Area | 135.00 | sq km | City Development Plan Bhubaneswar |
| Density | 4,741.00 | population/sq km | |
| Climate zone | Warm humid | – | National Building Code 2005 |
| Settlement type | Administrative | – | |
| Regional connectivity/ gateway status | 6 | – | Indian Railways, Airport Authority of India, Port Trusts of India, Inland Water Authority of India |
| Spatial structure | Linear 2 | – | Google Earth 2015 |
| Land use <br> • Industrial <br> • Transportation <br> • Parks and green areas | <br> 1.15 <br> 4.67 <br> 8.74 | <br> % of total urban area | City Development Plan Bhubaneswar |
| *Emissions, energy and waste data* | | | |
| Total emissions 2007 | 0.97 | Mt $CO_2$e | ICLEI 2009 |
| Per capita emissions | 0.71 | t $CO_2$e/year | ICLEI 2009 |
| Electricity <br> • Residential <br> • Commercial <br> • Industrial | <br> 323.60 <br> 318.26 <br> 64.73 | <br> M KWh | ICLEI 2009 |
| Annual fuel consumption | 99,141.00 | Kilolitres (KL) | ICLEI 2009 |
| MSW | 375.00 | Tons per day (tpd) | ICLEI 2009 |

*City image*: Sujit kumar/Wikipedia Commons/GNU Free Documentation License, Version 1.2

*Map*: Map data @2015 Google

### 5.2.5 Gwalior

Population – 820,000 (2001), 1,053,505 (2011)

| Field | Value | Unit | Source |
|---|---|---|---|
| *Spatial data* | | | |
| Area | 290.00 | sq km | City Development Plan Gwalior |
| Density | 2,832.00 | population/sq km | |
| Climate zone | Composite | – | National Building Code 2005 |
| Settlement type | Multifunctional | – | |
| Regional connectivity/ gateway status | 5 | – | Indian Railways, Airport Authority of India, Port Trusts of India, Inland Water Authority of India |
| Spatial structure | Organic 5 | – | Google Earth 2015 |
| Land use<br>• Industrial<br>• Transportation<br>• Parks and green areas | <br>6.28<br>17.06<br>3.62 | % of total urban area | City Development Plan Gwalior |
| *Emissions, energy and waste data* | | | |
| Total emissions 2007 | 0.49 | Mt CO$_2$e | ICLEI 2009 |
| Per capita emissions | 0.41 | t CO$_2$e/year | ICLEI 2009 |
| Electricity<br>• Residential<br>• Commercial<br>• Industrial | 253.30<br>105.40<br>28.50 | M KWh | ICLEI 2009 |
| Annual fuel consumption | 70,663.00 | Kilolitres (KL) | ICLEI 2009 |
| MSW | 285.00 | Tons per day (tpd) | ICLEI 2009 |

*City image*: YashiWong/Wikimedia Commons/GNU Free Documentation License, Version 1.2

*Map*: Map data @2015 Google

### 5.2.6 Madurai

Population – 920,000 (2001), 1,017,865 (2011)

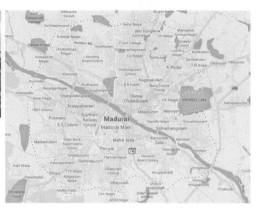

| Field | Value | Unit | Source |
|---|---|---|---|
| *Spatial data* | | | |
| Area | 109.00 | sq km | City Development Plan Madurai |
| Density | 8,440.00 | population/sq km | |
| Climate zone | Warm humid | – | National Building Code 2005 |
| Settlement type | Multifunctional | – | |
| Regional connectivity/ gateway status | 5 | – | Indian Railways, Airport Authority of India, Port Trusts of India, Inland Water Authority of India |
| Spatial structure | Radial 13 | – | Google Earth 2015 |
| Land use <br>• Industrial <br>• Transportation <br>• Parks and green areas | <br>6.00 <br>15.95 <br>7.96 | <br>% of total urban area | City Development Plan Madurai |
| *Emissions, energy and waste data* | | | |
| Total emissions 2007 | 0.28 | Mt CO$_2$e | ICLEI 2009 |
| Per capita emissions | 0.95 | t CO$_2$e/year | ICLEI 2009 |
| Electricity <br>• Residential <br>• Commercial <br>• Industrial | 50.00 <br>15.00 <br>9.00 | M KWh | ICLEI 2009 |
| Annual fuel consumption | 56,377.00 | Kilolitres (KL) | ICLEI 2009 |
| MSW | 450.00 | Tons per day (tpd) | ICLEI 2009 |

*City image*: Bernard Gagnon/Wikimedia Commons/GNU Free Documentation License, Version 1.2

*Map*: Map data @2015 Google

### 5.2.7 Mysore

Population – 750,000 (2001), 887,446 (2011)

| Field | Value | Unit | Source |
|---|---|---|---|
| *Spatial data* | | | |
| Area | 128.00 | sq km | City Development Plan Mysore |
| Density | 5,840.00 | population/sq km | |
| Climate zone | Temperate | – | National Building Code 2005 |
| Settlement type | Administrative | – | |
| Regional connectivity/ gateway status | 4 | – | Indian Railways, Airport Authority of India, Port Trusts of India, Inland Water Authority of India |
| Spatial structure | Radio-concentric 9 | – | Google Earth 2015 |
| Land use | | | City Development Plan Mysore |
| • Industrial | 13.48 | % of total urban area | |
| • Transportation | 16.10 | | |
| • Parks and green areas | 13.74 | | |
| *Emissions, energy and waste data* | | | |
| Total emissions 2007 | 0.94 | Mt $CO_2$e | ICLEI 2009 |
| Per capita emissions | 0.72 | t $CO_2$e/year | ICLEI 2009 |
| Electricity | | | ICLEI 2009 |
| • Residential | 237.70 | M KWh | |
| • Commercial | 92.23 | | |
| • Industrial | 380.38 | | |
| Annual fuel consumption | 81,800.00 | Kilolitres (KL) | ICLEI 2009 |
| MSW | 300.00 | Tons per day (tpd) | ICLEI 2009 |

*City image*: Ezhuttukari@ml.wikipedia/Creative Commons Attribution-Share Alike 3.0 Unported License

*Map*: Map data @2015 Google

### 5.2.8 Nashik

Population – 1,070,000 (2001), 1,152,326 (2011)

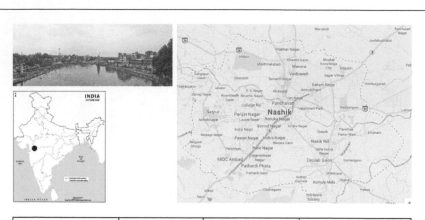

| Field | Value | Unit | Source |
|---|---|---|---|
| *Spatial data* | | | |
| Area | 259.00 | sq km | Development Plan of Nashik City |
| Density | 4,129.00 | population/sq km | |
| Climate zone | Hot dry | – | National Building Code 2005 |
| Settlement type | Multiple | – | |
| Regional connectivity/ gateway status | 4 | – | Indian Railways, Airport Authority of India, Port Trusts of India, Inland Water Authority of India |
| Spatial structure | Radial 9 | – | Google Earth 2015 |
| Land use<br>• Industrial<br>• Transportation<br>• Parks and green areas | <br>7.31<br>7.89<br>5.20 | % of total urban area | Development Plan of Nashik City |
| *Emissions, energy and waste data* | | | |
| Total emissions 2007 | 0.67 | Mt $CO_2$e | ICLEI 2009 |
| Per capita emissions | 0.91 | t $CO_2$e/year | ICLEI 2009 |
| Electricity<br>• Residential<br>• Commercial<br>• Industrial | 268.10<br>83.50<br>94.20 | M KWh | ICLEI 2009 |
| Annual fuel consumption | 1,011,146.00 | Kilolitres (KL) | ICLEI 2009 |
| MSW | 350.00 | Tons per day (tpd) | ICLEI 2009 |

*City image*: http://nashik.nic.in/index.html

*Map*: Map data @2015 Google

### 5.2.9 Patna

Population – 1,690,000 (2001), 2,046,652 (2011)

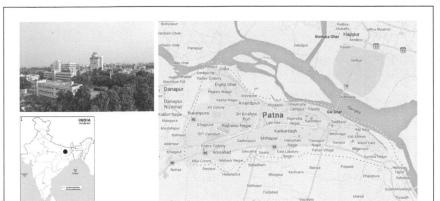

| Field | Value | Unit | Source |
|---|---|---|---|
| *Spatial data* | | | |
| Area | 135.00 | sq km | City Development Plan Patna |
| Density | 12,519.00 | population/sq km | |
| Climate zone | Composite | – | National Building Code 2005 |
| Settlement type | Administrative | – | |
| Regional connectivity/ gateway status | 6 | – | Indian Railways, Airport Authority of India, Port Trusts of India, Inland Water Authority of India |
| Spatial structure | Linear 3 | – | Google Earth 2015 |
| Land use<br>• Industrial<br>• Transportation<br>• Parks and green areas | <br>1.76<br>7.77<br>2.70 | % of total urban area | City Development Plan Patna |
| *Emissions, energy and waste data* | | | |
| Total emissions 2007 | 1.99 | Mt $CO_2$e | ICLEI 2009 |
| Per capita emissions | 1.32 | t $CO_2$e/year | ICLEI 2009 |
| Electricity<br>• Residential<br>• Commercial<br>• Industrial | <br>364.45<br>144.04<br>359.39 | M KWh | ICLEI 2009 |
| Annual fuel consumption | 330,156.00 | Kilolitres (KL) | ICLEI 2009 |
| MSW | 1,130.00 | Tons per day (tpd) | ICLEI 2009 |

*City image*: Nandanupadhyay/Creative Commons Attribution-Share Alike 3.0 Unported License
*Map*: Map data @2015 Google

### 5.2.10 Raipur

Population – 750,000 (2001), 1,010,087 (2011)

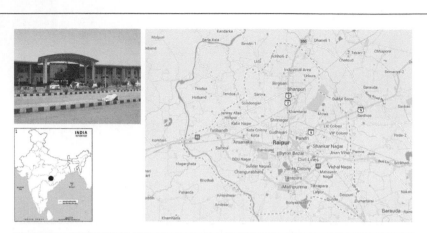

| Field | Value | Unit | Source |
|---|---|---|---|
| *Spatial data* | | | |
| Area | 154.00 | sq km | City Development Plan Raipur |
| Density | 4,870.00 | population/sq km | |
| Climate zone | Composite | – | National Building Code 2005 |
| Settlement type | Administrative | – | |
| Regional connectivity/ gateway status | 5 | – | Indian Railways, Airport Authority of India, Port Trusts of India, Inland Water Authority of India |
| Spatial structure | Organic 7 | – | Google Earth 2015 |
| Land use <br> • Industrial <br> • Transportation <br> • Parks and green areas | <br> 11.60 <br> 13.50 <br> 2.70 | <br> % of total urban area | City Development Plan Raipur |
| *Emissions, energy and waste data* | | | |
| Total emissions 2007 | 1.22 | Mt CO$_2$e | ICLEI 2009 |
| Per capita emissions | 0.37 | t CO$_2$e/year | ICLEI 2009 |
| Electricity <br> • Residential <br> • Commercial <br> • Industrial | <br> 263.58 <br> 90.92 <br> 50.49 | <br> M KWh | ICLEI 2009 |
| Annual fuel consumption | 207,073.90 | Kilolitres (KL) | ICLEI 2009 |
| MSW | 300.00 | Tons per day (tpd) | ICLEI 2009 |

*City image*: AshishJaiswalRaipur/Wikimedia Commons/GNU Free Documentation License, Version 1.2

*Map*: Map data @2015 Google

### 5.2.11 Shimla

Population – 140,000 (2001), 169,578 (2011)

| Field | Value | Unit | Source |
|---|---|---|---|
| *Spatial data* | | | |
| Area | 20.00 | sq km | Development Plan for Shimla, Town and Country Planning Department, GoHP |
| Density | 7,161.00 | population/ sq km | |
| Climate zone | Cold | – | National Building Code 2005 |
| Settlement type | Multifunctional | – | |
| Regional connectivity/ gateway status | 3 | – | Indian Railways, Airport Authority of India, Port Trusts of India, Inland Water Authority of India |
| Spatial structure | Linear 3 | – | Google Earth 2015 |
| Land use | | | Development Plan for Shimla, Town and Country Planning Department, GoHP |
| • Industrial | 0.62 | % of total | |
| • Transportation | 25.20 | urban | |
| • Parks and green areas | 0.41 | area | |
| *Emissions, energy and waste data* | | | |
| Total emissions 2007 | 0.14 | Mt $CO_2$e | ICLEI 2009 |
| Per capita emissions | 0.66 | t $CO_2$e/year | ICLEI 2009 |
| Electricity | | | ICLEI 2009 |
| • Residential | 53.92 | M KWh | |
| • Commercial | 25.30 | | |
| • Industrial | 1.53 | | |
| Annual fuel consumption | 17,727.00 | Kilolitres (KL) | ICLEI 2009 |
| MSW | 93.00 | Tons per day (tpd) | ICLEI 2009 |

*City image*: ShashankSharma2511/Wikimedia Commons/GNU Free Documentation License, Version 1.2

*Map*: Map data @2015 Google

### *5.2.12 Surat*

Population – 2,811,614 (2001), 4,461,026 (2011)

| Field | Value | Unit | Source |
|---|---|---|---|
| *Spatial data* | | | |
| Area | 327.00 | sq km | City Development Plan Surat |
| Density | 11,271.00 | population/sq km | |
| Climate zone | Hot dry | – | National Building Code 2005 |
| Settlement type | Multifunctional, Industrial | – | |
| Regional connectivity/ gateway status | 7 | – | Indian Railways, Airport Authority of India, Port Trusts of India, Inland Water Authority of India |
| Spatial structure | Radio-concentric 8 | – | Google Earth 2015 |
| Land use <br> • Industrial <br> • Transportation <br> • Parks and green areas | <br>17.74 <br>9.18 <br>0.62 | <br>% of total urban area | City Development Plan Surat |
| *Emissions, energy and waste data* | | | |
| Total emissions 2007 | 3.28 | Mt CO$_2$e | ICLEI 2009 |
| Per capita emissions | 1.20 | t CO$_2$e/year | ICLEI 2009 |
| Electricity <br> • Residential <br> • Commercial <br> • Industrial | 531.00 <br>414.00 <br>2,033.00 | M KWh | ICLEI 2009 |
| Annual fuel consumption | 228,508.00 | Kilolitres (KL) | ICLEI 2009 |
| MSW | 1,093.00 | Tons per day (tpd) | ICLEI 2009 |

*City image*: Vimal Chand/Wikpedia/Creative Commons

*Map*: Map data @2015 Google

### 5.2.13 Thiruvananthapuram

Population – 740,000 (2001), 889,635 (2011)

| Field | Value | Unit | Source |
|---|---|---|---|
| *Spatial data* | | | |
| Area | 142.00 | sq km | City Development Plan Thiruvananthapuram |
| Density | 5,221.00 | population/sq km | |
| Climate zone | Warm humid | – | National Building Code 2005 |
| Settlement type | Administrative | – | |
| Regional connectivity/ gateway status | 6 | – | Indian Railways, Airport Authority of India, Port Trusts of India, Inland Water Authority of India |
| Spatial structure | Radial 6 | – | Google Earth 2015 |
| Land use • Industrial • Transportation • Parks and green areas | 2.00 6.50 3.00 | % of total urban area | City Development Plan Thiruvananthapuram |
| *Emissions, energy and waste data* | | | |
| Total emissions 2007 | 0.23 | Mt CO$_2$e | ICLEI 2009 |
| Per capita emissions | 0.84 | t CO$_2$e/year | ICLEI 2009 |
| Electricity • Residential • Commercial • Industrial | 193.98 110.31 18.17 | M KWh | ICLEI 2009 |
| Annual fuel consumption | NA | Kilolitres (KL) | ICLEI 2009 |
| MSW | 250.00 | Tons per day (tpd) | ICLEI 2009 |

*City image*: Dikkoos/Creative Commons Attribution-Share Alike 3.0 Unported

*Map*: Map data @2015 Google

### 5.2.14 Vijayawada

Population – 850,000 (2001), 1,039,518 (2011)

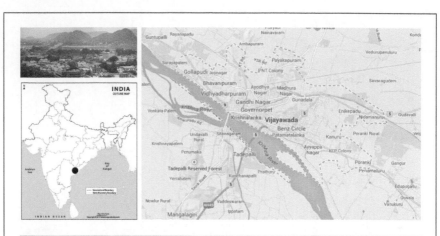

| Field | Value | Unit | Source |
|---|---|---|---|
| *Spatial data* | | | |
| Area | 110.00 | sq km | City Development Plan Vijayawada |
| Density | 7,717.00 | population/sq km | |
| Climate zone | Warm humid | – | National Building Code 2005 |
| Settlement type | Business/ commerce | – | |
| Regional connectivity/ gateway status | 6 | – | Indian Railways, Airport Authority of India, Port Trusts of India, Inland Water Authority of India |
| Spatial structure | Organic 5 | – | Google Earth 2015 |
| Land use • Industrial • Transportation • Parks and green areas | 2.95 20.95 3.79 | % of total urban area | City Development Plan Vijayawada |
| *Emissions, energy and waste data* | | | |
| Total emissions 2007 | 1.47 | Mt $CO_2$e | ICLEI 2009 |
| Per capita emissions | 0.91 | t $CO_2$e/year | ICLEI 2009 |
| Electricity • Residential • Commercial • Industrial | 398.88 190.72 162.36 | M KWh | ICLEI 2009 |
| Annual fuel consumption | 109,343.00 | Kilolitres (KL) | ICLEI 2009 |
| MSW | 350.00 | Tons per day (tpd) | ICLEI 2009 |

*City image*: Yedla 70/Creative Commons Attribution-Share Alike 3.0 Unported

*Map*: Map data @2015 Google

## 5.3 Spatial planning metric to estimate urban GHG emissions

In order to understand the functional relationship of the parameters that show the causal relationship towards GHG emissions, correlation and regression analysis is employed. The relationship could be represented by a simple mathematical equation:

$$Y = f(x_i)$$

Where:

$Y$ = GHG emissions/annum of a city, expressed either cumulatively or average per capita

$x_i$ = urban spatial parameter (climate type, regional connectivity, settlement scale, urban form, spatial structure, land use)

Correlation analysis for the selected cities (see Figure 5.1) shows that absolute or cumulative emissions of a city is a stronger indicator, in terms of coefficient of correlation, or $R^2$ (here, $R^2 = 0.8094$), across all the cities, than compared to its per capita emissions ($R^2 = 0.5994$). As such, the carbon footprint of a city can be better explained or represented through its overall emissions. Nonetheless, all USP are analyzed against total GHG emissions and per-capita GHG emissions and results discussed hereafter. The summary of results is reported in Table 5.4.

### 5.3.1 Climate type

*Results for (X1) climate type*

Sampled cities from diverse climate types are arranged in Table 5.1 and Figure 5.2a according to their "total residential and commercial energy consumption (in M KWh)" as a proxy indicator.

The correlation analysis shows a very strong positive correlation ($Y = 6E\text{-}08x^2 + 0.003x - 0.2323$, $R^2 = 0.937$) between the selected cities and their GHG emissions, as arranged in the following generic pattern:

| | cold < | temperate < | warm humid < | composite < | hot dry |
|---|---|---|---|---|---|
| Total residential and commercial energy consumption (M KWh): | 79 | 330 | 400 | 441 | 1,193 |

| | cold < | warm humid < | temperate < | composite < | hot dry |
|---|---|---|---|---|---|
| Average GHG emissions (Mt CO$_2$e): | 0.14 | 0.74 | 0.94 | 1.09 | 3.61 |

The results indicate that cities located in hot dry and composite regions are the most energy-consuming and carbon-intensive, owing to greater use of artificial means for air cooling/air conditioning. They also reveal that cold or temperate cities that typically consume energy mostly for heating needs are less carbon-intensive than cities located in hot dry, composite or warm humid locations that rather depend on air conditioning. The functional relation suggests that if the energy consumption for warm humid, temperate and composite locations is four to five times that of cold, then the GHG associated with it is five to eight times.

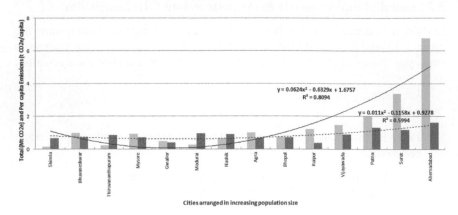

*Figure 5.1* Correlation of absolute volume (on left) and per capita emissions (on right) with different city size

*Table 5.1* Location, total and per capita GHGs, energy consumption of sample cities

| Name | Climate zone (NBC classification) | Total emissions (Mt $CO_2e$) 2007 | Per capita emmissions (t $CO_2e$/year) | Residential and commercial electricity consumption (M Kw) |
|---|---|---|---|---|
| Agra | Composite | 1.02 | 0.67 | 529 |
| Ahmedabad | Hot Dry | 6.78 | 1.63 | 2,282 |
| Bhopal | Composite | 0.74 | 0.71 | 453 |
| Bhuvaneshwar | Warm humid | 0.97 | 0.71 | 642 |
| Gwalior | Composite | 0.49 | 0.41 | 359 |
| Madurai | Warm humid | 0.28 | 0.95 | 65 |
| Mysore | Temperate | 0.94 | 0.72 | 330 |
| Nashik | Hot dry | 0.67 | 0.91 | 352 |
| Patna | Composite | 1.99 | 1.32 | 508 |
| Raipur | Composite | 1.22 | 0.37 | 355 |
| Shimla | Cold | 0.14 | 0.66 | 79 |
| Surat | Hot dry | 3.38 | 1.20 | 945 |
| Thiruvananthapuram | Warm humid | 0.23 | 0.84 | 304 |
| Vijayawada | Warm humid | 1.47 | 0.91 | 590 |

Similarly, in case of hot dry locations, as energy consumption is 15 times that of cold, the GHG associated with it is 26 times. The above results could be used to guide future strategies for low-carbon urbanization and industrialization. As evident, cities located in cold, temperate and warm humid conditions are rather conducive to sustainable and climate-friendly urban development.

When the same sample of cities is plotted for their increasing non-industrial (residential and commercial) energy consumption (M KWh) against the per capita GHG emissions from diverse climate types, as in Figure 5.2b, the analysis shows a moderate but positive correlation ($Y = 6E-08x^2 + 0.0003x + 0.6538$, $R^2 = 0.506$). The following pattern emerges:

| | cold | < temperate | < warm humid | < composite | < hot dry |
|---|---|---|---|---|---|
| Total residential and commercial energy consumption: (M KWh): | 79 | 330 | 400 | 441 | 1,193 |

| | cold | < temperate | < warm humid | < composite | < hot dry |
|---|---|---|---|---|---|
| Average GHG emissions/capita (t $CO_2$e/capita): | 0.66 | 0.72 | 0.85 | 0.99 | 1.24 |

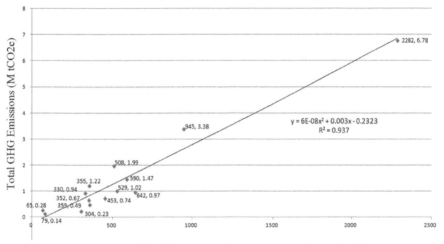

Total Residential and Commercial energy consumption (M KWh):

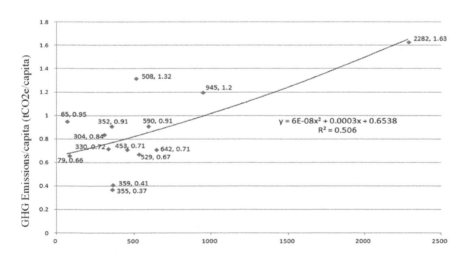

Total Residential and Commercial energy consumption (M KWh):

*Figure 5.2* (a) Correlation of total GHG emissions and residential and energy consumption. (b) Correlation of per capita GHG emissions and residential and energy consumption

The functional relation suggests that if the per capita residential and commercial energy consumption for warm humid, temperate and composite locations is four times that of cold, then the GHGs associated with it are 1.1 times. Similarly, in the case of hot dry locations, as energy consumption is 16 times that of cold, the GHG associated with it becomes double.

### *5.3.2 Regional connectivity*

*Results for (X2) number of linkages*

Sampled cities from diverse regional connectivity are arranged according to their increasing number of regional linkages – land-, water- and air-based in Figure 5.3a. Each linkage is counted as one (1), as explained in section 4.4. If the number of linkages increases, enhancing the gateway status of that place, say on account of an international airport or seaport, an additional point is attributed. Table 5.2 gives details for the number of linkages of the selected 14 cities and their respective total and per capita emissions.

The analysis shows a reasonably strong and positive correlation ($Y = 0.5476x^2 - 4.5982x + 9.7963$, $R^2 = 0.6772$) between the number of regional linkages and their GHG emissions, and could be arranged in the following generic pattern:

| Number of linkages: | 3 | 4 | 5 | 6 | 7 |
|---|---|---|---|---|---|
| Average GHG emissions (Mt $CO_2$e): | 0.14 | 0.71 | 0.78 | 1.16 | 5.08 |
| | | (+407.1%) | (+9.8%) | (+48.7%) | (+337.9%) |

*Table 5.2* Regional connectivity, total and per capita GHGs of sample cities

| Name | Number of regional linkages | Total emissions (Mt $CO_2$e) 2007 | Per capita emissions (t $CO_2$e/year) |
|---|---|---|---|
| Agra | 5 | 1.02 | 0.67 |
| Ahmedabad | 7 | 6.78 | 1.63 |
| Bhopal | 5 | 0.74 | 0.71 |
| Bhuvaneshwar | 6 | 0.97 | 0.71 |
| Gwalior | 4 | 0.49 | 0.41 |
| Madurai | 5 | 0.28 | 0.95 |
| Mysore | 4 | 0.94 | 0.72 |
| Nashik | 5 | 0.67 | 0.91 |
| Patna | 6 | 1.99 | 1.32 |
| Raipur | 5 | 1.22 | 0.37 |
| Shimla | 3 | 0.14 | 0.66 |
| Surat | 7 | 3.38 | 1.2 |
| Thiruvananthapuram | 6 | 0.23 | 0.84 |
| Vijayawada | 6 | 1.47 | 0.91 |

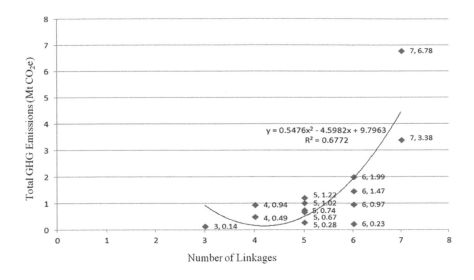

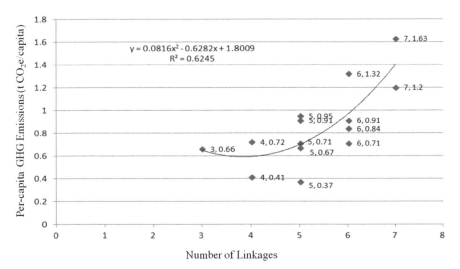

*Figure 5.3* (a) Correlation of total GHG emissions and number of linkages.
(b) Correlation of per capita GHG emissions and number of linkages

The results reveal that the greater the number of regional linkages of a city, the higher the prospect of its GHG emissions. Settlements with rail junctions, ports and airports bearing connectivity within the region or interregions (six linkages) or beyond, such as international networks (seven linkages), show a sudden rise in their carbon footprint, as in the case of Ahmedabad (6.78 Mt $CO_2$e) and Surat

(3.38 Mt $CO_2$e). It is observed that the carbon footprint of a city with an international airport or seaport is over four times that with domestic/regional gateway status. Thus, it needs to be noted that global connectivity through sea- or air-based modes comes at a significantly high carbon price, which may also have comparably high environmental costs in terms of air pollution, water scarcity, traffic congestion, solid waste management, etc.

When the same sample of cities is plotted for their increasing number of land-, water- and air-based linkages against the per capita GHG emissions, as evident in Figure 5.3b, the analysis shows a moderate but positive correlation ($Y = 0.0816x^2 - 0.6282x + 1.8009$, $R^2 = 0.6245$):

| Average number of linkages: | 3 | 4 | 5 | 6 | 7 |
|---|---|---|---|---|---|
| Average per capita GHG emissions (t $CO_2$e/capita): | 0.66 | 0.56 (−15.1%) | 0.72 (+28.5%) | 0.94 (+30.5%) | 1.41 (+50.0%) |

Analysis from Figure 5.3b demonstrates that, in general, per capita emissions of a city follow a similar trend to that of its total GHG emissions (with an exception of Shimla city, with three regional linkages and per capita emissions of 0.66). An increase in each regional linkage is accompanied by a 28–50% increase (a factor of 1.3–1.5) in carbon footprint when cities start having international linkages. Shimla shows abruptly higher per capita emissions than the average of Mysore (0.72 t $CO_2$e/capita) and Gwalior (0.41 t $CO_2$e/capita), probably because, being a hill city, it has a comparatively higher energy consumption per capita (0.83 M KWh) than Gwalior (0.47 M KWh).

### 5.3.3 Settlement scale

*Settlement scale studied with two indicators: (X3) urban area and (X4) urban population*

*Results for (X3) urban area*

Sampled cities of different scale/size are arranged according to their increasing urban (municipal) areas in Figure 5.4a. Analysis shows strong, positive correlation ($Y = 5E\text{-}05x^2 - 0.014x + 1.5543$, $R^2 = 0.7662$) between urban area and total GHG emissions, and could be arranged by following a generic pattern of results:

| Urban/municipal area (in sq km): | 20–466 |
|---|---|
| Total GHG emissions (Mt $CO_2$e): | 0.14–6.78 |

It could be generalized that the physical footprint of a city influences its carbon footprint. As the city expands and sprawls on land, so do the GHG emissions swell in the atmospheric space. While urban area multiplies by a factor of 23, the GHG do so by 49. It is vital to understand the sensitivity of per capita emissions when the city size increases. When the same sample of cities is plotted for their

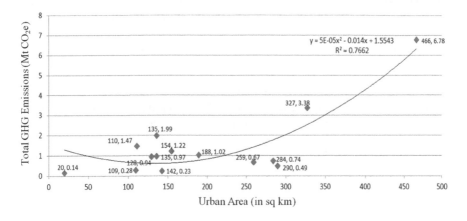

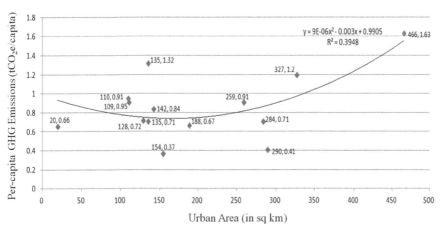

*Figure 5.4* (a) Correlation of total GHG emissions and urban area. (b) Correlation of per capita GHG emissions and urban area

increasing municipal areas against respective per capita GHG emissions, the analysis in Figure 5.4b shows positive but little correlation ($Y = 9E\text{-}06x^2 - 0.003x + 0.9905$, $R^2 = 0.3948$).

| Urban/municipal area (in sq km): | 20–466 |
|---|---|
| Per capita GHG emissions (t $CO_2$e/capita): | 0.66–1.63 |

### Results for (X4) urban population

Sampled cities of different scale/size are arranged in Figure 5.5a according to their increasing urban population (that of the municipal area or urban agglomeration, as applicable). The analysis shows a very strong and positive correlation

($Y = 1E\text{-}13x^2 + 3E\text{-}07x + 0.371$, $R^2 = 0.9436$) between a city's urban area and total GHG emissions. The results exhibit the following pattern:

| | |
|---|---|
| Urban/population: | 140,000–5,500,000 |
| Total GHG emissions (Mt $CO_2$e): | 0.14–6.78 |

The rise in urban population and GHG emissions is quite even, such that an approximate 40-fold amplification in a city's population is accompanied by an almost equal enlargement of its carbon footprint. It could be generalized that the population of a city influences its carbon footprint considerably. As the city population grows, so does the likelihood of its GHG emissions. It is important

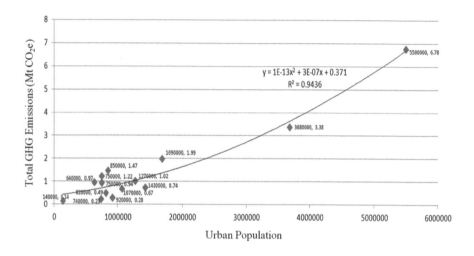

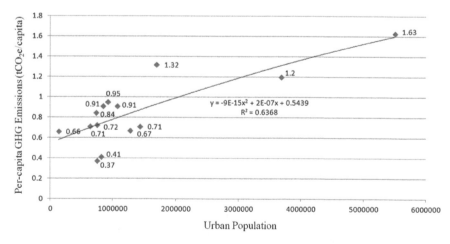

*Figure 5.5* (a) Correlation of total GHG emissions and urban population. (b) Correlation of per capita GHG emissions and urban population

to understand how sensitive population of a city is to its per capita emissions, as city size increases. When the same sample of cities is plotted for their increasing population against respective per capita GHG emissions in Figure 5.5b, the analysis shows a positive and moderately strong correlation ($Y = -9E\text{-}15x^2 + 2E\text{-}07x + 0.5439$, $R^2 = 0.6368$). One would assume that as cities grow in population, there is a greater prospect of the per capita emissions to reduce. The results suggest that if a city's population increases from 1 million to 5.5 million, its per capita emissions change from 0.6 to about 1.6 t $CO_2$e/capita only. This is quite similar to results from international cities that show economies of scale, owing to agglomeration. The carbon savings are on account of lesser energy and fuel used, transmission and distribution losses, etc. in provision of utilities, services and transport:

| | |
|---|---|
| Urban/population: | 140,000–5,500,000 |
| Per capita GHG emissions (t $CO_2$e/capita): | 0.66–1.63 |

### 5.3.4 Urban form

*Results for (X5) urban density*

Sampled cities of different form are arranged in Figure 5.6a according to their increasing urban densities. The analysis shows a moderate and positive correlation ($Y = 6E\text{-}08x^2 - 0.0005x + 1.8569$, $R^2 = 0.5431$) between their urban density and total GHG emissions, and could be explained through the following generic pattern:

| | |
|---|---|
| Urban density (in population per sq km): | 2,832–12,519 |
| Average GHG emissions (Mt $CO_2$e): | 0.49–6.78 |

As urban density increases from 2,832 per sq km (Gwalior) to 12,519 per sq km (Patna), the city's GHG emissions rise from 0.49 Mt $CO_2$e to 6.78 Mt $CO_2$e, respectively. The data suggest that a fourfold urban density is accompanied by an almost 14-fold carbon footprint. But no generalities could be derived, because there are many cities that have higher densities, such as Thiruvananthapuram (5,221/sq km), Shimla (7,161/sq km) and Madurai (8,440/sq km), but their overall carbon footprints are smaller. When the same sample of cities is plotted in Figure 5.6b for their increasing densities against respective per capita GHG emissions, the analysis shows a strong and positive correlation ($Y = 6E\text{-}09x^2 - 6E\text{-}06x + 0.5254$, $R^2 = 0.7584$):

| | |
|---|---|
| Urban density (in population per sq km): | 2,832–12,519 |
| Total GHG emissions (t $CO_2$e/capita): | 0.37–1.63 |

It emerges that as urban density increases from 2,832 per sq km (Gwalior) to 12,519 per sq km (Patna), the city's per capita GHG emissions rise from

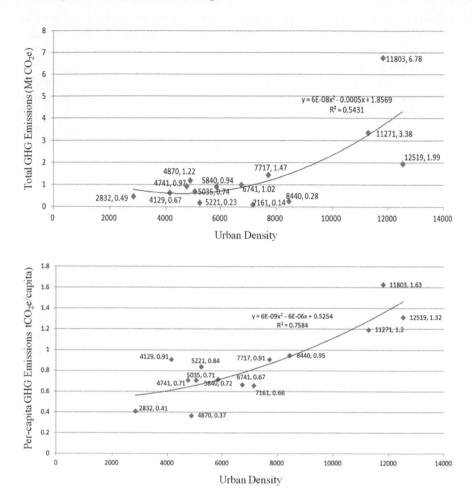

*Figure 5.6* (a) Correlation of total GHG emissions and urban density. (b) Correlation of per capita GHG emissions and urban density

0.37 (Raipur) to 1.63 (Ahmedabad) t $CO_2$e/capita. The rise in urban density and GHG emissions is quite even, such that an approximate fourfold amplification in a city's density is accompanied by a similar enlargement of its carbon footprint. It is important to view this function in comparison to other global results.

### 5.3.5 Spatial structure

*Results for (X6) city structure and (X7) number of arteries*

Sampled cities of different city structure and arteries (see Table 5.3) are arranged according to their diverse spatial structures in Figure 5.7a. The analysis shows a very strong and positive correlation ($Y = 0.0624x^2 - 0.6329x + 1.6757$, $R^2 = 0.8094$) between a city's spatial structure and its total GHG emissions, and could be depicted in the following arrangement:

| Spatial structure: | Radial | < Linear | < Organic | < RC | < RCC |
|---|---|---|---|---|---|
| Average GHG emissions (Mt $CO_2$e): | 0.55 | 1.03 | 1.06 | 1.68 | 6.78 |

*Table 5.3* Spatial structure, total and per capita GHGs of sample cities

| Name | Spatial structure | Total emissions (Mt $CO_2$e) 2007 | Per capita emissions (t $CO_2$e/year) |
|---|---|---|---|
| Agra | R 8 | 1.02 | 0.67 |
| Ahemadabad | RCC 16 | 6.78 | 1.63 |
| Bhopal | RC 7 | 0.74 | 0.71 |
| Bhuvaneshwar | L 2 | 0.97 | 0.71 |
| Gwalior | O 5 | 0.49 | 0.41 |
| Madurai | R 13 | 0.28 | 0.95 |
| Mysore | RC 9 | 0.94 | 0.72 |
| Nashik | R 9 | 0.67 | 0.91 |
| Patna | L 3 | 1.99 | 1.32 |
| Raipur | O 7 | 1.22 | 0.37 |
| Shimla | L 3 | 0.14 | 0.66 |
| Surat | RC 8 | 3.38 | 1.2 |
| Thiruvananthapuram | R 6 | 0.23 | 0.84 |
| Vijayawada | O 5 | 1.47 | 0.91 |

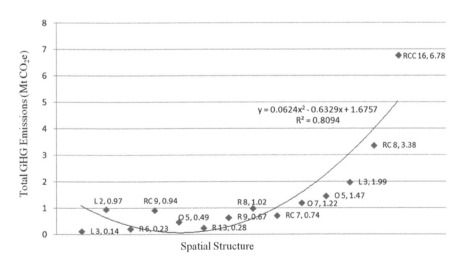

*(continued)*

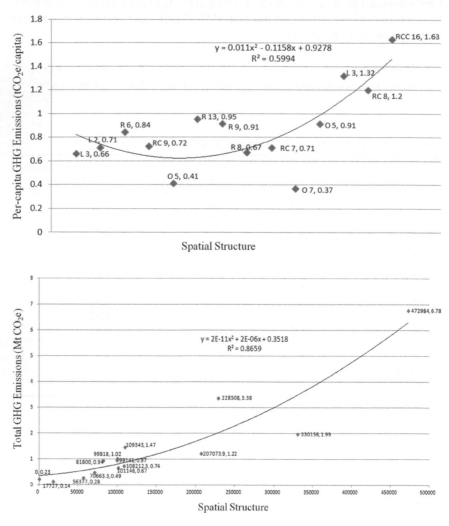

*Figure 5.7* (a) Correlation of total GHG emissions and spatial structure. (b) Correlation
of per capita GHG emissions and spatial structure. (c) Correlation of total
GHG emissions and annual fuel consumption

It reveals that on an average, radial cities (denoted by R) are least carbon intensive,
while complex radio-concentric (RCC) spatial structures with multi-nuclei economic function are most carbon intensive. Cities with linear (L) and organic (O)
spatial structures have twice as large carbon footprints than radial cities. Meanwhile
simple radio-centric (RC) cities with mono-centric economic hubs have carbon
footprints 3 times larger than radial cities. Similarly, complex radio-concentric
structures have carbon footprints 14 times as simple radial cities or 4 times that

of simple radio-concentric cities. Similarly, complex radio-concentrics have carbon footprints 14 times that of simple radials or four times that of simple radio-concentrics. When the same sample of cities is plotted in Figure 5.7b against per capita GHG emissions, the analysis shows a moderate and positive correlation ($Y = 0.011x^2 - 0.1158x + 0.9278$, $R^2 = 0.5994$). The following pattern emerges:

| Spatial structure: | Organic < Radial < RC | < Linear < RCC |
|---|---|---|
| Average per capita GHG emissions (t $CO_2$e/capita): | 0.56   0.84   0.87 | 0.89   1.63 |

Due to a lesser $R^2$ value and greater dispersion of results, it is prudent to avoid making any generalizations. When the same sample is analysed using annual transportation fuel consumption as a proxy indicator in Figure 5.7c, it emerges that there is a very strong and positive correlation between annual transportation fuel used within a city and its total GHG emissions ($Y = 2E\text{-}11x^2 + 2E\text{-}06x + 0.3518$, $R^2 = 0.8659$). The following pattern emerges:

| Spatial structure: | Radial < organic < RC < linear < RCC |
|---|---|
| Average annual fuel consumption (in KL): | 85,780   129,027   139,507   149,008   472,984 |

It is observed that cities, when arranged in increasing order of their annual transportation fuel consumption, in general follow the same pattern to that of GHG emissions, with linear cities (such as Bhuvaneshwar, Patna and Shimla) the only exception, and thus needing further investigation.

### 5.3.6 Land use

*Results for (X9) industrial land use*

Sampled cities of different land use are arranged according to their increasing area under industrial land use in Figure 5.8a. The analysis shows a reasonably strong and positive correlation ($Y = 0.0301x^2 - 0.3434x + 1.4101$, $R^2 = 0.6813$) between a city's industrial land use and its total GHG emissions, and could be depicted in the following arrangement:

| Area under industrial land use (% of total municipal/built-up area): | 0.62–17.74 |
|---|---|
| Total GHG emissions (Mt $CO_2$e): | 0.14–6.78 |

It is observed that, on average, as area under industrial land use within a city increases from about 1% to 17% of the total urban or developed area, its overall emissions multiply from 1.0 to 4.25 Mt $CO_2$e. Hence, every 1 percentage point increase in industrial land use within the city is accompanied by about 0.2 Mt $CO_2$e in its emissions. When the same sample of cities is plotted against per capita GHG emissions in Figure 5.8b, the analysis shows a rather positive but poor correlation ($Y = 0.0066x^2 - 0.1008x + 1.0359$, $R^2 = 0.4175$). Under such circumstances, any generalization of results could not be drawn in clear terms:

| | |
|---|---|
| Area under industrial land use (% of total municipal/built-up area): | 0.62–17.74 |
| Per capita GHG emissions (t $CO_2$e/capita): | 0.37–1.63 |

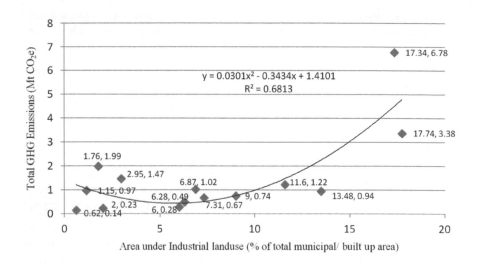

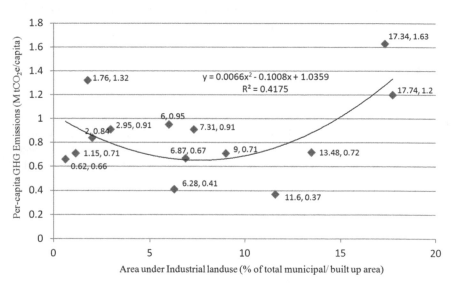

*Figure 5.8* (a) Correlation of total GHG emissions and area under industrial land use.
(b) Correlation of per capita GHG emissions and area under industrial land use

*Results for (X10) transportation land use*

Sampled cities of different land use are arranged according to their increasing area under transportation land use in Figure 5.9a. The analysis shows a very poor significance and negative correlation ($Y = 0.0109x^2 - 0.4895x + 5.8792$, $R^2 = 0.2876$)

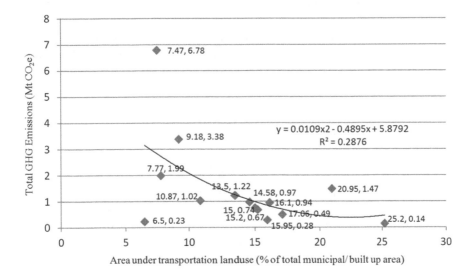

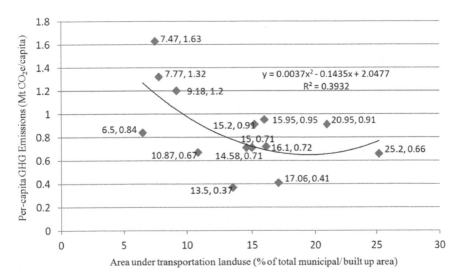

*Figure 5.9* (a) Correlation of total GHG emissions and area under transportation land use. (b) Correlation of per capita GHG emissions and area under transportation land use

between a city's transportation land use and its total GHG emissions, and could be depicted in the following arrangement. Under such poor correlation, it is difficult to derive any definite findings:

Area under transportation land use (% of total municipal/built-up area):      3–25
Total GHG emissions (Mt $CO_2$e):      0.14–6.78

When the same sample of cities is plotted against per capita GHG emissions in Figure 5.9b, the analysis yet again shows a poor significance and negative correlation ($Y = 0.0037x^2 - 0.1435x + 2.0477$, $R^2 = 0.3932$). The results do not suggest any significant correlation of values:

Area under transportation land use (% of total municipal/built-up area):      3–25
Per capita GHG emissions (t $CO_2$e/ capita):      0.37–1.63

*Results for (X11) recreational or green land use*

Sampled cities of different land use are arranged according to their increasing area under recreational or green land use in Figure 5.10a. The analysis shows a very poor significance and no definite correlation ($Y = -0.0159x^2 + 0.184x + 1.2599$, $R^2 = 0.0427$) between a city's recreational land use and its total GHG emissions, and could be depicted with the following arrangement:

Area under recreational land use (% of total municipal/built-up area):      0.41–13.74
Total GHG emissions (Mt $CO_2$e):      0.14–6.78

When the same sample of cities is plotted against per capita GHG emissions in Figure 5.10b, the analysis yet again shows a poor significance and non-definite correlation ($Y = -0.0033x^2 + 0.036x + 0.8131$, $R^2 = 0.0403$). With such findings, it is pointless to derive any major correlation of variables:

Area under recreational land use (% of total municipal/built-up area):      0.41–13.74
Per capita GHG emissions (t $CO_2$e/capita):      0.37–1.63

**Failure of results using linear, logarithmic, exponential and higher polynomial equations to establish correlation:** It is noteworthy to mention that while different types of functional equations were tried to find correlations between dependent (cumulative emissions and per capita emissions) and independent variables (X1 to X10), such as linear, logarithmic, exponential and higher polynomial equations, they all failed to give the best results for $R^2$ (coefficient of correlation). All independent variables show higher values of $R^2$ to the dependent variables using quadratic equations (see Table 5.4 for summary).

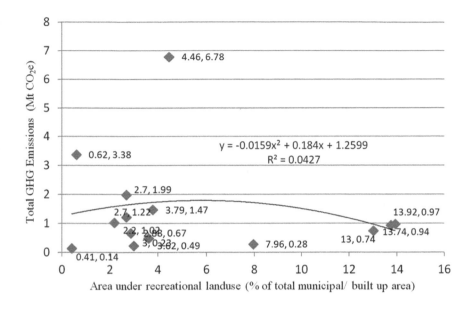

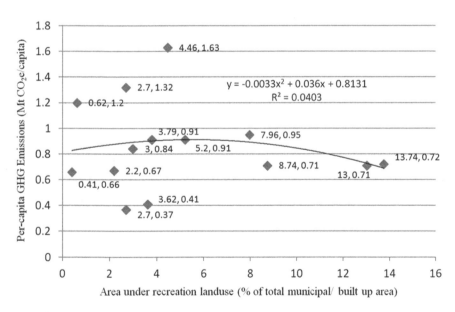

*Figure 5.10* (a) Correlation of total GHG emissions and area under recreational land use. (b) Correlation of per capita GHG emissions and area under recreational land use

*Table 5.4* Mathematical function/equation and $R^2$ values (coefficient of correlation/goodness of fit)

| Indicators | | Equation/function and $R^2$ | |
| --- | --- | --- | --- |
| | | Cumulative emissions (Mt $CO_2e$) | Per capita emissions ($CO_2e$/capita) |
| X1 | Climate type | $Y = 6E\text{-}08x^2 + 0.003x - 0.2323$, $R^2 = 0.937$ | $Y = 6E\text{-}08x^2 + 0.0003x + 0.6538$, $R^2 = 0.506$ |
| X2 | Number of linkages | $Y = 0.5476x^2 - 4.5982x + 9.7963$, $R^2 = 0.6772$ | $Y = 0.0816x^2 - 0.6282x + 1.8009$, $R^2 = 0.6245$ |
| X3 | Urban area | $Y = 5E\text{-}05x^2 - 0.014x + 1.5543$, $R^2 = 0.7662$ | $Y = 9E\text{-}06x^2 - 0.003x + 0.9905$, $R^2 = 0.3948$ |
| X4 | Urban population | $Y = 1E\text{-}13x^2 + 3E\text{-}07x + 0.371$, $R^2 = 0.9436$ | $Y = -9E\text{-}15x^2 + 2E\text{-}07x + 0.5439$, $R^2 = 0.6368$ |
| X5 | Urban density | $Y = 6E\text{-}08x^2 - 0.0005x + 1.8569$, $R^2 = 0.5431$ | $Y = 6E\text{-}09x^2 - 6E\text{-}06x + 0.5254$, $R^2 = 0.7584$ |
| X6 | City structure | $Y = 0.0624x^2 - 0.6329x + 1.6757$, $R^2 = 0.8094$ | $Y = 0.011x^2 - 0.1158x + 0.9278$, $R^2 = 0.5994$ |
| X7 | Number of arteries | | |
| X8 | Industrial land use | $Y = 0.0301x^2 - 0.3434x + 1.4101$, $R^2 = 0.6813$ | $Y = 0.0066x^2 - 0.1008x + 1.0359$, $R^2 = 0.4175$ |
| X9 | Transportation land use | $Y = 0.0109x^2 - 0.4895x + 5.8792$, $R^2 = 0.2876$ | $Y = 0.0037x^2 - 0.1435x + 2.0477$, $R^2 = 0.3932$ |
| X10 | Recreational land use | $Y = -0.0159x^2 + 0.184x + 1.2599$, $R^2 = 0.0427$ | $Y = -0.0033x^2 + 0.036x + 0.8131$, $R^2 = 0.0403$ |

Note: Variables with significant correlations ($R^2$ values greater than 0.6) are shaded.

## 5.4 Formulation of model

### 5.4.1 OLS regression model: description, application in urban/environmental/climate change studies, interpretation of results

Regression analysis is a statistical method to estimate the relationships among variables. It includes many techniques for modelling and analysing several variables, when the focus is on the relationship between a dependent variable and one or more independent variables. More specifically, regression analysis helps one understand how the typical value of the dependent variable (or 'criterion variable') changes when any one of the independent variables is varied, while the other independent variables are held fixed.

Regression analysis is also used to understand which among the independent variables are related to the dependent variable, and to explore the forms of these relationships. In restricted circumstances, regression analysis can be used to infer causal relationships between the independent and dependent variables.

**Ordinary least squares (OLS)**, or linear least squares, is a method for estimating the unknown parameters in a linear regression model. This method minimizes the sum of squared vertical distances between the observed responses in the data set and the responses predicted by the linear approximation. The resulting estimator can be expressed by a simple formula, especially in the case of a single regressor on the right-hand side. When only one dependent variable is being modelled, a scatter diagram will suggest the form and strength of the relationship between the dependent variable and regressors. It might also reveal outliers, heteroscedasticity and other aspects of the data that may complicate the interpretation of a fitted regression model. The scatter diagram suggests that the relationship is strong and can be approximated as a quadratic function. OLS can handle non-linear relationships by introducing the regressor area. The regression model then becomes a multiple linear model:

$$w_i = \beta_1 + \beta_2 h_i + \beta_3 h_i^2 + \varepsilon_i$$

In multiple linear regression, there are several independent variables or functions of independent variables. Adding the term $x_i^2$ to the preceding regression gives:

$$y_i = \beta_0 + \beta_1 x_i + \beta_2 x_i^2 + \varepsilon_i$$
$$i = 1 \ldots\ldots n$$

This is still linear regression. Although the expression on the right-hand side is quadratic in the independent variable $x_i$, it is linear in the parameters $\beta_0, \beta_1$ and $\beta_2$. A typical equation would exhibit a curve, with positive or negative curvature. Table 5.5 interprets the causal relationships of the selected parameters with GHGs based on positive and negative signs.

### 5.4.2 Data sheet of 41 cities

Table 5.6 presents the data sheet of 41 cities for selected parameters.

*Table 5.5* Interpretation of causal relationships – positive and negative

|  | Total emissions (Mt $CO_2e$) | Per capita emissions (t $CO_2e$/year) |
|---|---|---|
| **Non-industrial electricity (M KWh)** Proxy for climate type | + | + |
| **Population (Census 2001)** | + | + |
| **Regional connectivity** | + | + |
| **Area (sq km)** | + | + |
| **Density (population/sq km)** | + | + |
| **Industrial electricity (M KWh)** Proxy for industrial land use | + | + |
| **Annual fuel consumption (KL)** Proxy for city structure, number of arteries and transportation land use | + | + |

Table 5.6 Data sheet of 41 cities

| City | Independent variables | | | | | | | | Dependent variables | |
|---|---|---|---|---|---|---|---|---|---|---|
| | Population (Census 2001) | Regional connectivity | Area (sq km) | Density (population/ sq km) | Non-industrial electricity (M KWh) | Industrial electricity (M KWh) | Annual fuel consumption (KL) | Municipal solid waste (tpd) | Total emissions (Mt CO$_2$e) | Per capita emissions (t CO$_2$e/ year) |
| 1 Agra | 1,270,000 | 5 | 188 | 6,741 | 529 | 53.09 | 99,818 | 710 | 1.02 | 0.67 |
| 2 Ahemadabad | 5,500,000 | 7 | 466 | 11,803 | 2,282 | 2,266.62 | 472,984 | 2,242 | 6.78 | 1.63 |
| 3 Asansol | 470,000 | 4 | 127 | 3,694 | 87 | 9.95 | 32,340 | | 0.23 | 0.25 |
| 4 Bengaluru | 4,300,000 | 6 | 224 | 19,140 | 5,490 | 2,610.10 | 310,100 | 5,033 | 6.36 | 0.82 |
| 5 Bhavnagar | 510,000 | 5 | 53 | 9,568 | 338 | 79.45 | 133,909 | | 0.83 | 0.76 |
| 6 Bhopal | 1,430,000 | 5 | 284 | 5,035 | 453 | 29.5 | 108,212.30 | 550 | 0.74 | 0.71 |
| 7 Bhuvaneshwar | 640,000 | 6 | 135 | 4,741 | 642 | 64.73 | 99,141 | 375 | 0.97 | 0.71 |
| 8 Chennai | 4,340,000 | 7 | 181 | 23,970 | 765 | 78 | 525,150 | 3,641 | 3.82 | 2.25 |
| 9 Coimbatore | 930,000 | 5 | 105 | 8815 | 1022 | 144.19 | 152,622 | 601 | 1.27 | 1.15 |
| 10 Dehradun | 420,000 | 2 | 300 | 1,400 | 427 | 19.34 | 86,892 | 143 | 0.57 | 0.31 |
| 11 Faridabad | 1,050,000 | 2 | 208 | 5,048 | 1,341 | 756 | 343,158 | 480 | 2.46 | 0.67 |
| 12 Guntur | 510,000 | 2 | 63 | 8,076 | 505 | 93.75 | 45,900 | 356 | 0.56 | 0.31 |
| 13 Gurgaon | 170,000 | 2 | 120 | 1,417 | 1,202 | 503.28 | 476,712 | 570 | 2.55 | 0.3 |
| 14 Gwalior | 820,000 | 4 | 289 | 2,832 | 359 | 28.50 | 70,663.30 | 285 | 0.49 | 0.41 |
| 15 Haldia | 170,000 | 4 | 228 | 744 | 27 | 2.95 | 27,720 | 50 | 0.21 | 0.25 |
| 16 Indore | 1,470,000 | 5 | 214 | 6,869 | 625 | 117.03 | 141,472.33 | 600 | 1.14 | 0.83 |
| 17 Jabalpur | 930,000 | 5 | 154 | 6,031 | 271 | 18.8 | 81,695.10 | 330 | 0.46 | 0.64 |
| 18 Jaipur | 1,400,000 | 4 | 200 | 6,986 | 1,423 | 740.24 | 283,105 | 621 | 2.41 | 1.11 |
| 19 Jamshedpur | 570,000 | 5 | 230 | 2,472 | 1,204 | 3,765.20 | 247,679 | 560 | 5.51 | 0.33 |
| 20 Kanpur | 2,550,000 | 5 | 300 | 8,500 | 1,021 | 320.66 | 94,473 | 1,200 | 1.95 | 2.13 |
| 21 Kochi | 590,000 | 6 | 94 | 6,218 | 361 | 35.59 | – | 250 | 0.26 | 0.40 |
| 22 Kolkata | 4,570,000 | 7 | 185 | 24,703 | 2,181 | 503.16 | 606,942 | 4,000 | 9.33 | 2.76 |

| 23 | Lucknow | 2,180,000 | 5 | 310 | 7,032 | 2,040 | 372 | 142,402 | 1,550 | 2.37 | 1.97 |
| 24 | Madurai | 920,000 | 5 | 109 | 8,440 | 65 | 9 | 56,377 | 450 | 0.28 | 0.95 |
| 25 | Mysore | 750,000 | 4 | 128 | 5,840 | 330 | 380.38 | 81,800 | 300 | 0.94 | 0.72 |
| 26 | Nagpur | 2,050,000 | 6 | 218 | 9,404 | 690 | 479 | 171,263 | 770 | 1.65 | 1.58 |
| 27 | Nashik | 1,070,000 | 5 | 259 | 4,129 | 352 | 94.2 | 101,146 | 350 | 0.67 | 0.91 |
| 28 | Patna | 1,690,000 | 6 | 135 | 12,519 | 508 | 359.39 | 330,156 | 1,130 | 1.99 | 1.32 |
| 29 | Pune | 2,530,000 | 6 | 450 | 5,614 | 1,874 | 2,526.10 | 637,746 | 1,200 | 6 | 1.31 |
| 30 | Raipur | 750,000 | 5 | 154 | 4,870 | 355 | 50.49 | 207,073.90 | 300 | 1.22 | 0.37 |
| 31 | Rajkot | 960,000 | 4 | 104 | 9,155 | 450 | 360.49 | 30,098 | 227 | 0.88 | 0.64 |
| 32 | Ranchi | 840,000 | 4 | 111 | 7,568 | 389 | 200.13 | 334,969 | 360 | 2.88 | 0.45 |
| 33 | Sangli | 440,000 | 1 | 142 | 3,099 | 95 | 43.14 | 87,989 | 190 | 0.47 | 0.52 |
| 34 | Shimla | 140,000 | 3 | 19 | 7,161 | 79 | 1.53 | 17,727 | 93 | 0.14 | 0.66 |
| 35 | Surat | 3,680,000 | 7 | 326 | 11,271 | 945 | 2,033 | 228,508 | 1,093 | 3.38 | 1.2 |
| 36 | Thane | 1,260,000 | 3 | 147 | 8,571 | 686 | 537.73 | 39,458.72 | 600 | 1.45 | 1.83 |
| 37 | Thiruvananthapuram | 740,000 | 6 | 141 | 5,221 | 304 | 18.17 | – | 250 | 0.23 | 0.84 |
| 38 | Tiruchirapalli | 750,000 | 5 | 146 | 5,106 | 534 | 14 | – | 432 | 0.35 | 0.90 |
| 39 | Udaipur | 380,000 | 2 | 37 | 10,270 | 214 | 131.26 | 126,072 | 125 | 0.62 | 0.34 |
| 40 | Vijaywada | 850,000 | 6 | 110 | 7,717 | 590 | 162.36 | 109,343 | 350 | 1.47 | 0.91 |
| 41 | Vishakhapatnam | 980,000 | 7 | 550 | 1,782 | 700 | 982.69 | 195,257 | 880 | 7.36 | 1.37 |

## 5.5 Data findings

Data for the above 41 cities were input into the OLS regression model and the results were obtained for how two sets of dependent variable (i.e. (a) cumulative city emissions and (b) per capita city emissions) behave, which is discussed below.

### 5.5.1 Cumulative/total emissions of cities

---

## Box 5.1    OLS regression results for total emissions of 41 cities

Number of obs = 41

$F(6, 34) = 37.48$

Prob $> F = 0.0000$

$R$-squared = 0.8077

Root MSE = 1.0787

| $CO_2$_ total | Coef. | Robust Std. Err. | t | P>\|t\| | [95% Conf. Interval] | |
|---|---|---|---|---|---|---|
| popln | 5.45e-08 | 2.00e-07 | 0.27 | 0.787 | −3.52e-07 | 4.62e-07 |
| density | .0000116 | .0000724 | 0.16 | 0.874 | −.0001355 | .0001586 |
| reg_ conn | .2660214 | .2261024 | 1.18 | 0.248 | −.193474 | .7255168 |
| ind_elec | .0007513 | .0002134 | 3.52 | 0.001 | .0003177 | .001185 |
| fuel | 6.26e-06 | 1.65e-06 | 3.78 | 0.001 | 2.90e-06 | 9.62e-06 |
| nonind_ elec | .0003969 | .0002045 | 1.94 | 0.061 | −.0000188 | .0008126 |
| _cons | −1.193013 | .7488504 | −1.59 | 0.120 | −2.71486 | .3288337 |

**Mathematical Equation:**

$Y$ (total GHG emissions of a city) = 5.45e-08 (population)

+ 0.0000116 (density)

+ 0.2660214 (regional connectivity)

+ 0.0007513 (consumption of industrial electricity in M KWh)

+ 6.26e-06 (annual fuel consumption in KL)

+ 0.0003969 (non-industrial electricity consumption in M KWh)

− 1.193013

---

### 5.5.2 Per capita emissions of cities

---

## Box 5.2 OLS regression results for per capita emissions of 41 cities

Number of obs = 41

$F(6,34) = 13.35$

Prob $> F = 0.0000$

$R$-squared = 0.6199

Root MSE = .40428

| $co_2\_$ total | Coef. | Robust Std. Err. | t | $P>|t|$ | [95% Conf. Interval] | |
|---|---|---|---|---|---|---|
| **popln** | 3.06e-07 | 1.32e-07 | 2.33 | 0.026 | 3.90e-08 | 5.74e-07 |
| **density** | .0000131 | .0000225 | 0.58 | 0.562 | −.0000325 | .0000588 |
| **reg_ conn** | .0598767 | .0547919 | 1.09 | 0.282 | −.0514738 | .1712272 |
| **ind_ elec** | −.0001598 | .0001101 | −1.45 | 0.156 | −.0003835 | .000064 |
| **fuel** | 3.34e-07 | 5.82e-07 | 0.57 | 0.570 | −8.49e-07 | 1.52e-06 |
| **nonind_ elec** | −.0000652 | .0001596 | −0.41 | 0.685 | −.0003895 | .000259 |
| **_cons** | .1960053 | .240567 | 0.81 | 0.421 | −.2928857 | .6848962 |

**Mathematical Equation:**

$Y$ (per capita GHG emissions of a city) = 3.06e-07 (population)

+ 0.0000131 (density)

+ 0.0598767 (regional connectivity)

+ 0.0001598 (consumption of industrial electricity in M KWh)

+ 3.34e-07 (annual fuel consumption in KL)

− 0.0000652 (non-industrial electricity consumption in M KWh)

+ 0.1960053

### 5.5.3 A summary of all variables

---

**Box 5.3   Observations, mean, standard deviation, minimum and maximum**

| Variable | Obs | Mean | Std. Dev. | Min | Max |
|---|---|---|---|---|---|
| co$_2$_pc | 41 | .9314634 | .6045559 | .25 | 2.76 |
| popln | 41 | 1,404,146 | 1,310,569 | 140,000 | 5,500,000 |
| density | 41 | 7,550.537 | 5,191.442 | 744 | 24,703 |
| reg_ conn | 41 | 4.707317 | 1.584991 | 1 | 7 |
| ind_elec | 41 | 512.0778 | 863.2679 | 1.53 | 3,765.2 |
| fuel | 41 | 512.0778 | 168,168.5 | 0 | 637,746 |
| nonind_ elec | 41 | 823.2927 | 944.0494 | 27 | 5,490 |

*Mutual correlation of all variables*

| | co$_2$_pc | popln | density | reg_conn | ind_elec | fuel | nonind_ elec | co$_2$_ total |
|---|---|---|---|---|---|---|---|---|
| co$_2$_pc | 1.0000 | | | | | | | |
| popln | 0.7291 | 1.0000 | | | | | | |
| density | 0.6565 | 0.7638 | 1.0000 | | | | | |
| reg_ conn | 0.5429 | 0.6116 | 0.4525 | 1.0000 | | | | |
| ind_elec | 0.1314 | 0.4737 | 0.1386 | 0.3041 | 1.0000 | | | |
| fuel | 0.4550 | 0.6426 | 0.4988 | 0.3095 | 0.5039 | 1.0000 | | |
| nonind_ elec | 0.3522 | 0.6752 | 0.4841 | 0.3015 | 0.6400 | 0.5353 | 1.0000 | |
| co$_2$_total | 0.5468 | 0.7109 | 0.4860 | 0.4977 | 0.7009 | 0.7879 | 0.6870 | 1.0000 |

---

The results for both sets of dependent variables, (a) total city emissions and (b) per capita city emissions, are reported with respect to the independent variables in Table 5.7.

The comparison suggests that, without any doubt whatsoever, the total GHG emissions of a city serve to be the most appropriate metric while representing its USPs. It can be depicted most effectively by the following mathematical function/ equation:

$$Y \text{ (total GHG emissions of a city)} = 5.45\text{e-}08\,(X4) + 0.0000116\,(X5) + 0.2660214\,(X2) + 0.0007513\,(X8) + 6.26\text{e-}06\,(X6) + 0.0003969\,(X1) - 1.193013$$

The findings show that industrial activity, settlement structure and climate tend to be critical parameters that define a city's emission footprint.

*Table 5.7* Results for both sets of dependent variables, (a) total city emissions and (b) per capita city emissions, with respect to the independent variables

| | Dependent variable 1: total emissions (Mt $CO_2e$) | Dependent variable 2: per capita emissions (t $CO_2e$/year) |
|---|---|---|
| | *Coeff P>\|t\|* | *Coeff P>\|t\|* |
| Population (Census 2001) | 5.45e-08 (0.787) | 3.06e-07 (0.026) |
| Density (population/sq km) | 0.0000116 (0.874) | 0.0000131 (0.562) |
| Regional connectivity | 0.2660214 (0.248) | 0.0598767 (0.282) |
| Industrial electricity (M KWh) Proxy for industrial land use | 0.0007513 (0.001)*** | 0.0001598 (0.156) |
| Annual fuel consumption (KL) Proxy for city structure, number of arteries and transportation land use | 6.26e-06 (0.001)** | 3.34e-07 (0.570) |
| Non-industrial electricity (M KWh) Proxy for climate type | 0.0003969 (0.061)* | −0.0000652 (0.685) |
| Constant | −1.193013 | +0.1960053 |
| *R*-square | 0.8077 | 0.6199 |
| No. of observations | 41 | 41 |

Note: Figures in parenthesis represent robust standard errors. ***, ** and * indicate statistical significance at the 1%, 5% and 10% levels, respectively.

The GHG emissions of activity are not significantly related to its population and population density. A 1,000% increase in density (i.e 10-fold) leads to a 0.01% increase in total GHG emissions of a city.

Whereas a 10% increase in regional connectivity of a city leads to a 2.66% increase in its GHG emissions. It could be implied that if there is a 100% increase (i.e. doubling of connections, say from 3 to 6), the increase in the city's GHGs would be a sizeable 26%.

A 1,000% (i.e. 10-fold) increase in industrial activity is associated with a 0.7% increase in GHG emissions, while an equal increase in non-industrial electricity in residential and commercial activities increases the GHG emissions by 0.4%.

___

## Basic questions

1 What are the criteria to select a sample of cities to collect data for a spatial matrix to account for GHG emissions?

2 What is correlation analysis? Discuss its merit while conducting a pilot study on a sample of cities to ascertain their GHG emission pattern.

3 Justify why the metric of total GHG emissions of a city is a superior indicator in representing its carbon footprint than GHG emission per capita.

4   What is OLS regression and how is it beneficial in multivariate analysis?
5   What is a dependent and independent variable in correlation and regression analysis?
6   What is a typical mathematical equation to represent GHG emissions of a city based on its urban spatial parameters?

## Do it yourself exercises

1   Based upon the data inventory used in this chapter (section 5.2), prepare posters on GHG emission inventories of megacities in the world. Compare and deliberate your results.
2   Using the typical equation of total GHG emissions of a city based on its USP (given at the end of section 5.5.3), evaluate the carbon footprint of your city. Alternatively, a group can do this activity for various cities of your country.

## References

ICLEI (2009). *Energy and carbon emissions profiles of 54 South Asian cities.* New Delhi: International Council for Local Environmental Initiatives.

National Building Code (2005). *National Building Code of India. Second reprint.* New Delhi: Bureau of Indian Standards.

# 6 Research and policy implications

Science can only ascertain what is, but not what should be, and outside of its domain value judgements of all kinds remain necessary.

(Einstein, 1941)

## 6.1 Conclusions

This research, a first of its kind, reveals that in light of economic development, growing urbanization and GHG emissions, the relevance of cities and their governance cannot be undermined. In response to the four main research gaps identified within the inter-disciplines of cities and climate change (identified in Chapter 1) (i.e. conceptual gap, methodological gap, empirical gap and policy-governance gap), the conclusions and its ramifications, recommendations thereof could be broadly interpreted into the following four streams, namely: (1) scientific knowledge; (2) low-carbon urbanization policy; (3) planning and design: tools, methodologies and index; and (4) urban management and governance. These are diagrammatically shown in Figure 6.1 and are elaborated further.

### 6.1.1 Scientific knowledge

This research brings 'urban spatial' to the forefront, with new evidence at multiple scales. It corroborates that at the global scale, urbanization of countries is an important indicator that strongly associates with their GHG emissions, as against the normative economic indicators such as GDP. Even if a country is economically less developed but urbanized, it is bound to have a larger carbon footprint. While disaggregating GHG emissions from spatial perspective (urban-rural disposition) at the scale of an individual nation, with India as the case in point, it is proven without any doubt that although urban settlements host about a third of country's population, their contribution to national GHG emission is substantial, ranging between 66.5–70.3%. A large share of emissions emanating from urban areas is due to thermal power plants, cement and iron and steel industries. The research informs how closely the growth trajectories of economic development, urbanization and GHG emissions are interrelated. It divulges emission burdens across different sized Indian cities (small towns to megacities) and how it could influence

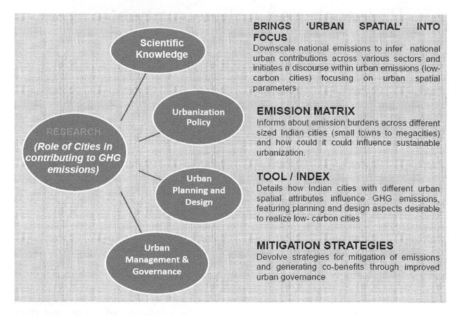

*Figure 6.1* The current research outcomes influence four areas/directions

sustainable urbanization. It shows that against the normal perception, Class I *census towns* (other than megacities) are the biggest emitters (57% of national emissions), followed by Class II–VI *census towns* (about 30% of national emissions). On the other hand, while megacities contributed to about 13% of India's urban population in 2011, their estimated contribution to urban GHGs is 5.65% only.

While studying local spatial parameters at an individual city level, the findings suggest that, all else being equal, it is the city's regional connectivity, industrial activities and spatial organization/structure that are critical to its GHG footprint. A city with large areas under industrial landuse, complex spatial structures like radio-concentric systems with greater number of arteries, and with adverse climate like acute hot-dry conditions show significant results for GHG emissions. On the contrary, conventionally used urban indicators like city population, urban density, regional linkages, etc. relatively show less statistical correlation with GHG emissions. The study also concludes with firm empirical evidence that, for a particular city, its cumulative or total emissions indicate or represent its carbon footprint more accurately than its per capita emission values.

### 6.1.2 Low-carbon urbanization policy

By studying the pattern of urbanization, economic growth and GHGs of world nations in the last five decades (1960–2010), the study underscores strong correlations between the state of urbanization and its emission footprint. The data of

209 countries suggest that the degree of national urbanization is associated directly with their GHG footprints, irrespective of that society's economic disposition. For example, many highly urbanized countries from less developing regions have larger carbon footprints than that of developed countries. Comparing this with their energy patterns, data from 209 countries suggest that uncontrolled urbanization could lead to a sudden increase in per capita energy intensity (due to greater consumption of fuels and electricity), and hence could lead to magnification of the carbon footprint. Hence, it is possible to attain a low-carbon pathway for a country through a rather planned and sustainable urbanization. While studying spatial parameters and their correlation with GHG emissions across sampled cities located in different geophysical or climate conditions, size/hierarchy, urban form and functional type, the following results (presented in Table 6.1) could be concluded for preferable/sustainable urbanization that bears minimal climate impacts.

### 6.1.3 Planning and design: tools, methodologies and index

Systematic modelling, simulation and optimization tools are an integral part of facilitating policy planning, design alternatives and the decision-making process in national and international affairs related to climate change, urbanization and energy access/provision. This research puts forward the application of the following three tools for systemic assessment of urban emissions at different levels:

1   *At the global/international scale*: spatial development matrix.
2   *At the national scale*: spatial disaggregation/downscaling emissions.
3   *At the settlement level*: correlation and regression (OLS)-based model.

*Table 6.1* Summary of results comparing urban spatial parameters and their GHG impacts

| Indicators | Favourable results (lower third of GHGs) | Mediocre results (middle third of GHGs) | Unfavourable results (upper third of GHGs) |
|---|---|---|---|
| | Low impacts | Medium impacts | High impacts |
| Climate type | Cold | Temperate, warm humid, composite | Hot, dry |
| Number of linkages | Less than or equal to 3 | 4–6 | 7 and above |
| Urban area | Less than or equal to 300 sq km | 300–400 sq km | Greater than 400 sq km |
| Urban population | Less than or equal to 2,500,000 | 2,500,000–4,000,000 | Greater than 400,000 |
| Urban density | 2,500–7,000 | 7,000–10,000 | Above 10,000 |
| Spatial structure | Radial | Linear, organic, radio-concentric (lower order/simple) | Radio-concentric (higher order/ complex) |
| Industrial land use | 0.6–11% | 11–14% | Above 14% |

At the international level, these tools could be used by different countries to assess the spatial perspective of their carbon footprint at multiple levels, and appropriately inform their urbanization policy. The spatial development matrix helps to plot country positions on the basis of their economic, urbanization and emissions/energy status, and provides a rationale for equitable, just and inclusive climate governance, access of carbon space, carbon credits and funds.

At the national level, these tools help assess the GHG contributions from urban areas and the main drivers. They further assist in identifying the potential climate co-benefits across different sectors. At the city level, the most important output is the functional relationship/mathematical equation that helps to estimate or correlate a city's total GHG emissions based on its spatial attributes. The entire technique conforms to the UNFCCC principles of TCCCA, and should invariably employ MRV methodologies. The process could thus equip urban local bodies (ULBs), their technical experts and decision-makers to take appropriate measures in plans, projects and day-to-day affairs that further help in mitigating city's emissions, thereby reducing its carbon footprint. City-level modelling also provides a basic framework or template for cities to optimize, simulate and compare their GHGs on the basis of their spatial attributes.

### 6.1.4 Urban management and governance

There are definite drawbacks in prevailing climate governance and ill preparedness of urban settlements to respond to changing climate. This research, the first of its kind, presents the baseline situation of climate change mitigation in an increasingly urbanizing India, bearing a strong emphasis to estimate and manage the GHG burden for appropriate urban governance. While urban areas contribute to almost two-thirds of national GDP and 70.3% of national GHG emissions, they stand with limited wherewithal, receiving merely 1.59% of the GDP as total expenditure on urban infrastructure.

Further, the capacities of ULBs are deeply imperilled on account of devolution of powers. While electricity generation from thermal power plants is the chief contributor of emissions (about 60%), followed by the cement and iron and steel industries (about 10% each), the ULBs have no mandate, or a restricted mandate, over these subjects. In spite of the clear directions of the 74th Constitutional Amendment Act (CAA) and recommendations made by various commissions, it is a known fact that states have acted little to provide autonomy to the ULBs to deal with sources or activities that contribute to GHG emissions, and hence influence climate change. The government's policy on the subject lacks sectoral integration, financing and market-based instruments to control GHG emissions. Mitigation of climate change was also absent in one of the biggest government investment and reform schemes in the urban sector – the JNNURM. Nonetheless, mitigation of climate change provides a good opportunity for cities to revisit their energy needs, regulations, provisions and enabling mechanisms, and also review waste management and land use/land cover management practices. In this regard, the government of India's latest flagship scheme in the urban sector (i.e. the Smart Cities Mission) addresses the challenge of energy-friendly urbanization head-on, though in select cities.

## 6.2 Recommendations for low-carbon city planning

Until now, low-carbon strategies in India were conceptualized from a totally sectoral perspective. It was implicit that reducing energy- and transportation-related GHG emissions are bound to reduce India's carbon footprint. Most often, strategies are focused on technological improvisations and resource conservation, both being driven in silos by several individual economic sectors. This study offers a spatial perspective to this equation, and infers that uncontrolled urbanization could lead to a sudden spurt in GHG emissions (above 6 t/capita). It is hence recommended that better results for a low-carbon society/economy could be achieved through sustainable urbanization, in the range of 34–67%, with less than 6 t/capita of average GHG emissions.

Integration of India's economic, energy and emission discourse with its urbanization policy is paramount. The policy should be based upon scientific and reasonable assessment of urban carbon footprints. From the application of multiscalar methodologies and tools to capture a spatial and holistic perspective of the carbon footprint, it is recommended that low-carbon urbanization should be kept as a concerted goal within the national urban development policy.

It is recommended that practical downscaling methods should be used to underpin urban contributions to national GHG burdens. This would provide long-term targets of urban GHGs for the national urban ministry, which could further be reported and monitored against policy actions. The reporting could start for metros, and later expand to all ULBs. The performance could be evaluated by incorporating certain indicators in existing service-level benchmarks. All the monitoring could be classified on a state basis, which would eventually influence technical, financial and capacity-building aspects. In addition, cities should inventorize their "urban spatial parameters" within the mandated master/city development plans as low-carbon policies or climate plans, set up benchmarks or measurable targets, and regularly monitor progress to achieve favourable results.

Based on the GHG impacts of various order (low, medium, high), any national goal directed towards sustainability, climate change, low-carbon/smart development, urbanization, industrialization, regional growth or spatial/environmental planning should identify or map geophysical areas that are conducive to urban development (can be termed as *go-areas*), conditionally conducive (*watch and go areas*) and non-conducive (*no-go areas*). For instance, hot-dry areas covering parts of Rajasthan, Gujarat and Maharashtra are no-go areas for highly industrialized and urbanized settlements, as far as carbon footprint is concerned. Accordingly, the government policies could be focused on urban containment in these areas. On the other hand, economic and urban policies could allow greater freedom of urbanization and expansion in relatively conducive areas.

A national low-carbon urbanization policy also needs to consider the optimal size of settlements. This could be either by limiting the urban area within 300 sq km or 2,500,000 as the maximum population. Similarly, towns that are already hosting higher populations could reduce their carbon impacts by aspiring to attain more compact and efficient urban forms, densifying within a range of 2,500–7,000 persons per sq km. It needs to be noted that against conventional wisdom, higher density may not always reduce GHG emissions, as it is usually accompanied by growing population size, thereby leading to higher GHG emissions.

Spatial organization or structure plays a significant role in limiting GHG impacts. Towns that generally develop from the simplest of villages or rural settlements tend to adopt a radial pattern, which happens to have a minimal GHG footprint. Thus, these naturally evolved or minimally planned structures are the most carbon-friendly. Linear, organic and first-order radio-concentric settlements also have lower carbon impacts. The structural order of a city offers a template to consolidate patterns of population density, its land use, urban form, movement and choice for transportation modes. It is pertinent for master plans, structure plans, city development plans, etc., as well as policymakers, urban planners and designers, to focus on sustainable patterns of spatial organization.

The study also concludes that among all urban land uses, industrial land use happens to be the most critical in affecting a city's GHG emissions. It goes without saying that industrial cities, particularly energy-intensive ones such as iron and steel, cement, manufacturing, etc., are highly carbon-emitting and should be planned and regulated in a suitable manner. But it is the multifunctional cities hosting GHG-emitting industrial activities that generally escape a strict urban policy and regulation. This research hence considered only multifunctional cities as samples, irrespective of the type of industry (scale, function, etc.) therein, and it could be inferred that industrial land use of less than 11% (as a percentage of total urban land use) produces the least environmental (emission) impacts. This seems in line with government guidelines for industrial development in urban–regional areas (UDPFI, 1996; URDPFI, 2014), and this value could be kept as the upper limit for industrial land use within a multifunctional city.

## 6.3 Mitigation-inclusive urban governance

Governance is "the quality of the relationship between the government and the citizens". Operationally, it is defined as "the quality of the process by which decisions are taken that affect public affairs, as well as the quality of the implementation and outcome of these decisions" (Sheng, 2010, p. 134). The mute question is how responsible urban governance can lead to realizing low-carbon city governance. In order to explore mitigation-inclusive urban governance, the most extensively mandated framework globally, by UN-Habitat (2011) and the OECD (2009), is followed. Its scope covers four basic modes of governance (i.e. through self-governance, provision, regulation and enabling across various sectors), and recommends for suitable mitigation mechanisms therein (as summarized in Table 6.2).

Table 6.2 Mitigation-inclusive urban governance framework in India – summary

| | Self-governing | Regulation | Provision | Enabling |
|---|---|---|---|---|
| **Energy generation** | • Energy audits and planning for all municipal services; energy saved is energy produced<br>• Installation of model WTE projects by ULBs<br>• Enhance purchased renewable energy<br>• Purchase of energy-efficient appliances and equipment in public infrastructure | • Setting up of cap-and-trade mechanism between states, where cities can also trade | • Transfer of electricity functions to ULBs or make distribution companies accountable<br>• Provision of smart grid service for consumers to link them to active demand–supply situation | • Incentives ought to be given to captive plant projects in renewable energy such as solar, wind, WTE, etc. that are to be installed for a ULB in city limits or nearby.<br>• Campaigns for energy conservation |
| **Transport** | • GIS and GPS systems to monitor all municipal vehicles<br>• Reduction of emissions from employees, through mobility plan, carpools, etc.<br>• Procurement of low-carbon-emitting vehicles – BS IV compliant for city fleet and municipal vehicles<br>• Demonstration projects to economically produce biofuel<br>• Use of low-carbon technology in road construction and maintenance overhaul | • Setting up of urban metropolitan authorities<br>• Planning and regulations for transit-oriented development<br>• Congestion charges and high parking price in congested areas<br>• Removal of tax and subsidies on fuels<br>• Oil companies to supply biofuels in municipal areas<br>• High taxation on personal vehicle registration, especially on inefficient or luxury models | • Greater percolation of mass rapid transport, especially electricity-based<br>• Provision of infrastructure for alternate or non-motorized modes such as cycle, walk, rickshaw, etc. | • Incentives, tax relaxations on use of hybrid and electric vehicles<br>• Awareness and education campaigns to use low-carbon or zero-emission transport modes<br>• Incentives on FAR for transit-oriented development projects |

(continued)

*Table 6.2* (continued)

| | Self-governing | Regulation | Provision | Enabling |
|---|---|---|---|---|
| **Residential and commercial** | • Switch to green buildings and having energy-efficient technologies in all construction works of the municipal government, both new and those involving renovation.<br>• Demonstration projects at house, neighbourhood, community scales for replication | • Revision of norms, standards for higher FAR, TDR in municipal core to encourage compact habitations and mixed land use<br>• Penalties or high property tax on vacant houses to curb speculative demand<br>• Mandatory application of ECBC throughout India<br>• All new residential and commercial developments to conform to green building guidelines | • Transfer of planning powers and provisionary functions to ULBs<br>• Heavy investments in rental housing, labour housing, studio apartments and industrial housing, which is low-cost and low-carbon footprint<br>• Provision of low-carbon municipal infrastructure | • Rebates on property taxes for green buildings, mixed use development<br>• Guidance to architects, engineers, developers on energy-efficient buildings<br>• Campaigns towards green buildings<br>• Ease of grants/loans for low-carbon technologies in households and businesses |
| **Industry** | • Low-carbon considerations in management of ULB's estates, procurements, etc. | • Transfer of regulatory functions of SPCB to ULBs or industrial estates in urban limits<br>• Cap-and-trade market with reduction targets for GHG emissions<br>• Polluter pays penalties on high emissions – chimney and captive plants/DG sets | • Phased overhaul of municipal infrastructure in industrial estates<br>• Provision of designated power supply to reduce load on captive (DG) generation | • Ease in grants/loans to industries for technology overhaul<br>• Enable list of energy auditors and firms<br>• Incentives of higher FAR, TDR for low emissions<br>• Rebates on property tax for consistent compliance |

|  | | | | |
|---|---|---|---|---|
| **Waste** | • Waste prevention, recycle, reuse within the ULB<br>• Procurement of low-carbon waste technologies and management practices<br>• Local solutions for waste segregation, composting and incineration<br>• Model projects for scientific landfill, methane recovery and WTE<br>• Use of GPS, GIS and MIS to measure and monitor waste generation, collection, transport, composting, dumping and incineration activities | • Ban on excessive packaging and plastic products<br>• Penalties on non-segregated waste at household or community level<br>• Regulation of municipal and e-waste through welfare associations and accredited junk dealers | • Recycling, recomposing, reuse schemes by ULBs<br>• Overhaul of waste water infrastructure and full coverage in municipal areas | • Campaigns for reusing, recycling, reducing wastes<br>• Promote use of recycled products<br>• Enabling greater role of private/PPP, NGOs and citizen groups in solid waste management<br>• Awareness campaigns against direct combustion of compostable waste |
| **LULUCF** | • Plantation in city parks, gardens, public places, city forests with self efforts or joint teams with forest department, NGOs, welfare associations, schoolchildren, etc. | • Planning and regulatory functions to revise master plan against horizontal sprawl and suburbanization<br>• Transfer of regulatory functions related to planning, change of land use, development controls to ULBs in municipal areas and peripheral areas<br>• Higher FAR to urban cores ensuring mixed land use<br>• Linking plantations with cap-and-trade market of industrial and power sector<br>• Setting up of minimum standards for keeping public parks and large private properties under green cover | • Devolution of management rights of parks, playgrounds and gardens to ward committees and welfare associations | • Incentives for peripheral properties to retain agricultural land use<br>• Awareness campaigns for enhancing tree cover |

Source: Author's recommendations drawn from Sethi and Mohapatra (2013); framework from UN-Habitat (2011) and OECD (2009) based on Bulkeley and Kern (2006), Bulkeley et al. (2009), Martinot et al. (2009) and ICLEI (2010)

Based on the findings of this research and the primary results drawn from Sethi and Mohapatra (2013), recommendations for mitigation-inclusive urban governance within respective sectors is presented henceforth.

**Electricity:** As Sethi (2015) suggests, about 90% of thermal emissions in India are produced within urban and rapidly urbanizing areas, and access to and use of clean energy in cities is imperative. Smart grids and power reforms that reduce generation and transmission losses can prove to become good governance mechanisms on the supply side. Meanwhile, it is noteworthy to evaluate the relevance of reducing in fossil fuel subsidies, taxes or carbon charges on fossil fuels, renewable energy obligations and loans on renewable energy production that are some international best practices. In growing ULBs, power generation functions from waste, etc. could be devolved to municipal governments under the ambit of the 74th CAA by respective states. This would bring more responsibility and accountability on the ground. Certain interventions on the demand side could prove instrumental in controlling consumption-side emissions. On this front, devolution of distribution functions to ULBs or regulatory powers can strengthen their position. Cap-and-trade mechanisms between states where ULBs also have stakes will trigger low-carbon technologies and practices, and complete the demand and supply gaps within the sector.

**Transport:** Urban local bodies play a significant role in the mitigation of carbon emissions arising from motor transport. In many Indian cities, there are sector-specific para-statal agencies, such as transport corporation/undertaking, and so on, to manage city traffic and bus services. Further emphasis is needful on mandatory fuel economy, bio-fuel blending and $CO_2$ standards for road transport, taxes on vehicle purchase, registration, motor fuels or roads, parking pricing, investment in public, and land use integrated transport. Integrated land use planning with transit corridors by assigning development hubs and allocating incentive zoning can result in a positive shift towards public transport, thereby reducing carbon emissions, as against indiscriminately permitting mixed land use along entire corridors, as proposed in several development plans, including the Master Plan of Delhi.

**Residential and commercial:** Since emissions in the residential and commercial sector are associated with broad energy needs for sustenance and socio-economic well-being, these are representative of the lifestyle and built environment an urban society lives in, and also how cities consciously choose to govern themselves. Analysis of selected Indian cities in this research reveals that as the municipal area expands from 20 to 466 sq km, the overall city emissions rise from 0.14 to over 6.78 Mt $CO_2$e, thereby showing a strong co-relation between city size and fuel emissions. Further, as city density increases from 2,832 to 12,519 per sq km, the carbon emissions rise from 0.49 to 6.78 Mt $CO_2$e, while per capita carbon emissions increase from 0.37 to 1.63 t $CO_2$e, thus showing costs and benefits on account of agglomeration and sharing of infrastructure need to be carefully examined against their GHG impacts. Offering higher FAR, tradable development rights and moderating property tax in inner city areas can result into counter-sprawl. Additionally, residential and commercial structures need to be made smarter through energy-efficient devices, green building practice, and waste management systems to reduce community and municipal emissions.

Residential and commercial emissions account for electricity consumption at the community level. The Indian government does recognize that, for good governance, municipal bodies should be encouraged to take responsibility of urban power distribution in their areas. This, however, should be done after adequate capacity building in these organizations (SARC, 2007). Further, municipal building by-laws should incorporate power conservation measures. In this regard, mandatory application of the Energy Conservation Building Code (ECBC) all over India is still in the offing. Meanwhile, Green Rating for Integrated Habitat Assessment (GRIHA), developed jointly by the Energy and Resources Institute and the Ministry of New and Renewable Energy, Government of India, is a green building "design evaluation system" and a crucial step in this direction. It is suitable for all kinds of buildings located in different climatic zones of the country, and aims to reduce over 30% energy, consume less water and limit waste to landfills (GRIHA, 2012).

**Industry:** Emissions from industry need to be dealt with by both production and consumption aspects. Global best practices suggest the provision of benchmark information, performance standards, subsidies, tax credits, tradable permits and voluntary agreements (World Bank, 2010). In 2010, the first ever Emission Trading System was initiated by Tokyo Metropolitan Government in Tokyo (PADECO report, cited in UN-Habitat, 2011), with the target to reduce $CO_2$ emissions by 25% below 2000 levels by 2020. It is recommended that, like power generation, it is high time that industries in urban areas are regulated by a cap-and-trade market mechanism with the government and citizen groups collectively acting as watchdogs. Measurement and monitoring of their assets by the ULBs will be a fundamental step in this direction. This would encourage low-carbon considerations in management of their estates, procurements, etc. Other initiatives would include implementing polluter pays principle by imposing penalties on high emissions – chimney and captive plants/DG sets, etc. In order to execute this, it is paramount to transfer regulatory functions of SPCB to ULBs or industrial estates. Certain specific measures for industries would go a long way in their enabling, such as an ease in grants/loans to industries for technology overhaul, incentives of higher FAR, TDR for low emissions, rebates on property tax for consistent compliance, and listing of energy auditors and firms.

**Waste:** Urban local bodies worldwide and in India have been constituted to provide basic services of sanitation. This has been strengthened by Municipal Solid Waste Rules 2000 under the Environmental Protection Act 1986, by which the responsibility of managing solid waste by scientific planning and management, setting up of landfill sites and treatment plants is delegated to the municipal bodies. The carbon reclaimed in the waste-to-energy practices, though small, will give fourfold benefits, by halving GHGs from wastes while doubling energy production from present levels, holding major opportunities for the future. It is suggested to follow the SARC (2007) recommendations for solid waste management issues in local governance, which mandates that: (a) for all towns and cities with a population above 0.1 million, the possibility of taking up PPP projects for collection and disposal of garbage may be explored, after creating suitable capacities to manage contracts; (b) municipal by-laws/rules should provide for segregation of waste into definite categories based on its manner of final disposal; and (c) special solid waste management charges should be levied on units generating a high amount of solid waste. Thus, a balanced enabling and regulatory approach seems most appropriate to the Indian case.

Some self-governing steps by the ULBs could be waste prevention, recycling, reusing within their organization, finding local solutions for waste segregation, composting and incineration, procurement of low-carbon waste technologies and management practices, starting model projects for scientific landfill, methane recovery, WTE, etc., and use of advanced systems to measure and monitor waste generation, collection, transport, composting, dumping and incineration activities. Regulatory governance includes a ban on excessive packaging/plastic products and regulation of municipal and e-waste through welfare associations and accredited junk dealers. As part of service provision function, governments should focus upon recycling, composting, reuse schemes by ULBs. In addition, overhaul of waste and wastewater infrastructure and achieving full coverage in municipal areas should be a priority. Enabling activities could make use of awareness campaigns in the society towards reusing, recycling and reducing wastes, promoting the use of recycled products, and facilitating private/PPP, NGOs and citizen groups in waste management.

**LULUCF:** The sector is generally undermined for urban forestry and agriculture, though many cities have intermittent yet significant open lands available. There exists vast potential to sequester $CO_2$ emissions within cities. For instance, in the recent past, green cover of Delhi has increased from 26 sq km in 1997 to about 296.20 sq km in 2009 (FSI, 2011) by the setting up of an eco joint force. The increasing agro-forestry products hold a multifold benefit as they enhance the mitigation capacities of the urban system and reduce the heat-island effect, thus limiting the use of air conditioning in buildings and vehicles, and also diminish the transportation emissions on account of the haulage of food and vegetables into cities from villages. The augmentation of the assimilative capacities to local carbon emissions creates additional benefits. As this research suggests, the increase in area under green land use within a city is negatively correlated to its total GHG emissions.

For a long time, statutory bodies involved in plan preparation for the city indulged in acquiring cheaply available lands for long-term planning process and there by driving suburbanization, horizontal sprawls and change in land-use, land cover. In the present urban milieu, there is no incentive for citizens to conserve their land for environmental provisions and build vertically, which would consequently reduce unnecessary travel time and allow them to enjoy a better lifestyle. As Indian cities grow from 100, 500, 1,000, 2,000 to 3,000 sq km, the average trip length increases from 3, 6.5, 7, 10 to 12 km, respectively (MoUD, 2008), indicating unnecessary sprawl. Hence, it is suggested to prevent and control the horizontal sprawl of cities.

Depending upon the city size, the policy for any new planned development seeks 12–25% of the developed area under recreation (UDPFI, 1996), and many planned cities actually have sufficient recreation areas with a multi-tier hierarchy of open spaces. As such, there is umpteen potential to mitigate potential carbon emissions through greens management, if exercised earnestly. Lower opportunities prevail in cities where the planning authorities were unable to meet with the development pace, and have let squatter settlements, slums and unplanned colonies come up, with no local parks and gardens. There is immense scope to redevelop such areas, opening them with breathing space on the ground by building vertically upwards. There is a lot of deliberation to actually make urban green management a conscious citywide effort, participatory and outcome-oriented. Additionally, all such functions need to be made financially sustainable, providing incentives such as higher FAR, and at the same time sourcing funds from taxes and cap-and-trade market mechanisms needful in the power and industry sectors. Thus, there is a far greater role to be played by ULBs in this sector through new regulations/provisions, enabling, etc., specifically by active land use planning.

Application of this research to mitigation-inclusive urban governance in India indicates that the steps to be taken vary drastically across sectors. They ought to be market-oriented with trading systems, where the government acts as the regulator only in the cases of industry, power and LULUCF. At the same time, interventions need to be authoritative and provision-based (along with the private sector and civil society organizations), as in the cases of the transport and waste sectors. In certain other situations, a blend of restricting and enabling mechanisms are required, as seen in the case of controlling indiscriminate urban sprawls. Yet, one aspect is common, that all sectors equally require strong communication, participation and incentives to stakeholders to move on low-carbon pathways. But, prior to outreach and collaborations, there are numerous initiatives to be taken by the government to set its house in order. The future of urban GHG emissions in India vastly depends on how well various levels of the federal system – centre, states and local bodies – respond to the call on climate change, exercise their specific powers in respective domains, and in chorus share collective responsibilities among themselves. It is an appropriate time that the ULBs stand up for innovative regulatory and market-based governance, while the states devolve larger city planning and management functions to these constitutional bodies and the national government focuses upon ensuring effective policy, mustering international support and financial commitments for local-level action.

## 6.4 Contribution to research, practice and further applications

The study is exemplary in multiple ways. There are limited studies across the globe that have tried to theoretically and empirically discern economic and carbon throughputs simultaneously on the urban–rural continuum, from the perspective of a physical/spatial planner. The challenge of this research could be gauged from the fact that this area of research is in its incipient stage in developing countries such as India and has limited data sets for many indicators, particularly at the city level, in comparison to developed countries. The research methodologically tackled this by focused analysis at multiple spatial scales, namely the global, national urban–rural and at the city level. The sampled cities were also carefully selected to represent the diversity of location, geophysical conditions, city size, etc. within the urban–rural hierarchy.

The research contributes to the theory with new findings, as summarized above. Some of these found acceptance and recognition as peer-reviewed journal papers and book chapters, before this volume of work. The research also inputs practice by providing new methods/tools for spatial planners to rationally and equitably account for carbon space, as well as by recommending an urbanization policy and urban governance framework inclusive of a green/low-carbon agenda.

Nonetheless, this research offers several opportunities for further and deeper investigations. These pertain to data, methods and outcomes/application, as detailed below:

- This study gives an essential base to further investigate the results in light of any new GHG accounting data released by the government, preferably at the subnational or local levels.
- Based on the definition and functional meaning of cities, particularly in India, this research utilizes production-/location-based methodologies to account urban GHG emissions. It would be imperative to further study city carbon footprints from the perspective of consumption pattern. At present, there is a paucity of data on resource consumption and economics at the local city level, which could lead to vague results. But a mix of survey and disaggregation methods may be employed to estimate consumption patterns of various goods and services within cities.
- At the local or city level, an index for urban spatial parameters could be prepared, specially for *census towns* of Class I and II size, that are undergoing higher levels of economic growth, population increase and urban expansion. The tools thus collectively inspire research into applications such as evaluation of planning alternatives, selecting the best possible option, its performance and monitoring. Meanwhile, the urban governance framework can be further expanded to include requirements for climate change adaptation.

---

## Basic questions

1    What role or control does economy of a country play within the relationship of its urbanization level and GHG emissions?
2    What are the favourable results for different USPs in causing low GHG impacts?
3    Which analytical tools could be applied for systemic assessment of urban GHGs at different levels of climate governance (global to city)?
4    What would an ideal low-carbon urbanization policy for your nation contain?

## Advanced questions

1    Deliberate on major spatial perspectives (or global–local implications) of how cities contribute to GHG emissions?

2    How could sustainable urbanization help attain a low-carbon pathway for a country?
3    How could spatial perspective of carbon footprint influence/guide regional and city planning policies? Demonstrate with examples from your country.
4    Examine whether urban governance and policy frameworks in your country are capable of mitigating GHG emissions. Elaborate.
5    What planning and governance instruments are needful to effectively mitigate subnational GHG emissions?
6    While studying the spatial perspective of the carbon footprint of urban areas, what are the research and application-oriented outcomes? Draw upon future research pathways.
7    For a particular GHG-contributing sector (transport/energy/LULUCF), discuss self-governing, regulatory, provisioning and enabling measures of governance to mitigate urban GHG emissions in your country.

## Do it yourself exercises

1    Using the analytical framework given in this chapter (see Table 6.2), prepare a poster on the urban governance framework of your country, evaluating all economic/GHG accounting sectors. This exercise could be done as a workshop too.
2    Drawing on the methods and findings of this research, discuss and finalize major points that could feed into a low-carbon urbanization policy of your country. Start with a broad wishlist and coalesce or synchronize issues to draw proposals that have multiple benefits. Conclude with four or five firm recommendations.

## Suggested reading

1    Bulkeley, H. and Betsill, M. (2003). *Cities and climate change: Urban sustainability and global environmental governance*. London: Routledge.
2    OECD (2009). *Cities, climate change and multilevel governance*. OECD Environmental Working Papers No 14. Paris: OECD Publishing.

## References

Bulkeley, H. and Kern, K. (2006). Local government and the governing of climate change in Germany and the UK. *Urban Studies*, *43*(12), 2237–59.
Bulkeley, H., Schroeder, H., Janda, K., Zhao, J., Armstrong, A., Chu, S.Y. and Ghosh, S. (2009). Cities and climate change: The role of institutions, governance and urban planning. Paper presented in the 5th Urban Research Symposium, *Cities and climate change*: *Responding to an urgent agenda*, 28–30 June, Marseille, France.
Einstein, A. (1941). Science, philosophy and religion: A symposium. Published by the *Conference on Science, Philosophy and Religion in Their Relation to the Democratic Way of Life, Inc.* New York, 1941. Cited in C. Seeling (Ed.), *Ideas and opinions by Albert Einstein* (5th edition, 1960). New York: Crown Publishers, p. 45. Retrieved from https://namnews.files.wordpress.com/2012/04/29289146-ideas-and-opinions-by-albert-einstein.pdf on 19 September 2016.

FSI (2011). *India Report*. New Delhi: Ministry of Environment and Forests, Government of India.

GRIHA (2012). *Green rating for integrated habitat assessment*. Retrieved from www. grihaindia.org/ on 18 September 2012.

ICLEI (2010) *Cities in a post-2012 climate policy framework: Climate financing for city development? Views from local governments, experts and businesses*. Retrieved from www.iclei.org/fileadmin/user_upload/documents/Global/Services/Cities_in_a_Post-2012_Policy_Framework-Climate_Financing_for_City_Development_ICLEI_2010. pdf on 7 August 2013.

Martinot, E., Zimmerman, M., Staden, M.V. and Yamashita, N. (2009). *Global status report on local renewable energy policies*. Retrieved from www.ren21.net/pdf/ REN21_LRE2009_Jun12.pdf on 9 May 2012.

MoUD (2008). *Study on traffic and transportation policies and strategies in urban areas in India*. New Delhi: Ministry of Urban Development, Government of India.

OECD (2009). *Cities, climate change and multilevel governance*. OECD Environmental Working Papers No 14. Paris: OECD Publishing.

SARC (2007) *6th Report: Local Governance*. New Delhi: Government of India.

Sethi, M. (2015). Location of greenhouse gases (GHG) emissions from thermal power plants in India along the urban–rural continuum. *Journal of Cleaner Production, 103*, 586–600. doi:10.1016/j.jclepro.2014.10.067.

Sethi, M. and Mohapatra, S. (2013). Governance framework to mitigate climate change: Challenges in urbanising India. In Huong Ha and Tek Nath Dhakal (Eds.), *Governance approaches to mitigation of and adaptation to climate change in Asia* (pp. 200–30). Hampshire, UK: Palgrave Macmillan.

Sheng, Y.P. (2010). Good urban governance in Southeast Asia. *Environment & Urbanisation Asia, 1*(2), 131–47.

UDPFI (1996). *Urban development plan and formulation guidelines India*. New Delhi: Ministry of Urban Development, Government of India.

URDPFI (2014). Urban and regional development plans formulation & implementation guidelines, Volume 1- first draft report. New Delhi: Ministry of Urban Development, Government of India.

UN-Habitat (2011). *Cities and climate change: Global report on human settlements*. London/Washington, DC: Earthscan & UNCHS.

World Bank (2010). *Climate resilient cities: A primer on reducing vulnerabilities to disasters*. Washington, DC: World Bank.

# Appendix I

# Summary tabulations

2010

| | Carbon dioxide emissions (*metric tons per capita*) | URBANIZATION LEVEL - Urban population (*percentage of total population*) URBAN $\longleftarrow \longrightarrow$ RURAL | | |
|---|---|---|---|---|
| | 2007 | **67% above** | **34-67%** | **≥ 33%** |
| **More Developed Regions** | 8.18 | 9.07 | 6.21 | 12.45 |
| **Less Developed Regions (excluding LDC)** | 5.51 | 8.3 | 3.47 | 4.41 |
| **Least Developed Countries** | 0.49 | 0.64 | 0.76 | 0.22 |

2000

| | Carbon dioxide emissions (*metric tons per capita*) | URBANIZATION LEVEL - Urban population (*percentage of total population*) URBAN $\longleftarrow \longrightarrow$ RURAL | | |
|---|---|---|---|---|
| | 2007 | **67% above** | **34-67%** | **≥ 33%** |
| **More Developed Regions** | 8.41 | 9.49 | 6.37 | 14.71 |
| **Less Developed Regions (excluding LDC)** | 5.15 | 9.66 | 3.08 | 2.58 |
| **Least Developed Countries** | 0.27 | 0.56 | 0.29 | 0.25 |

1990

| | Carbon dioxide emissions (*metric tons per capita*) | URBANIZATION LEVEL - Urban population (*percentage of total population*) URBAN $\longleftarrow \longrightarrow$ RURAL | | |
|---|---|---|---|---|
| | 2007 | **67% above** | **34-67%** | **≥ 33%** |
| **More Developed Regions** | 9.5 | 10.62 | 6.9 | 12.98 |
| **Less Developed Regions (excluding LDC)** | 4.67 | 8.3 | 3.51 | 1.82 |
| **Least Developed Countries** | 0.25 | 0.67 | 0.35 | 0.19 |

*(continued)*

*(continued)*

1980

|  | Carbon dioxide emissions (*metric tons per capita*) | URBANIZATION LEVEL - Urban population (*percentage of total population*) | | |
|  |  | URBAN ←———————————————→ RURAL | | |
|  | 2007 | **67% above** | **34-67%** | **≥ 33%** |
| **More Developed Regions** | 9.9 | 11.55 | 7.43 | 9.8 |
| **Less Developed Regions (excluding LDC)** | 5.14 | 13.84 | 3.73 | 1.44 |
| **Least Developed Countries** | 0.26 | 0.97 | 0.55 | 0.2 |

1970

|  | Carbon dioxide emissions (*metric tons per capita*) | URBANIZATION LEVEL - Urban population (*percentage of total population*) | | |
|  |  | URBAN ←———————————————→ RURAL | | |
|  | 2007 | **67% above** | **34-67%** | **≥ 33%** |
| **More Developed Regions** | 8.98 | 12.03 | 6.39 | 4.2 |
| **Less Developed Regions (excluding LDC)** | 5.12 | 16.69 | 4.25 | 1.01 |
| **Least Developed Countries** | 0.23 | 0 | 0.87 | 0.21 |

1960

|  | Carbon dioxide emissions (*metric tons per capita*) | URBANIZATION LEVEL - Urban population (*percentage of total population*) | | |
|  |  | URBAN ←———————————————→ RURAL | | |
|  | 2007 | **67% above** | **34-67%** | **≥ 33%** |
| **More Developed Regions** | 6.09 | 10.21 | 3.73 | 1.48 |
| **Less Developed Regions (excluding LDC)** | 1.47 | 4.13 | 1.67 | 0.39 |
| **Least Developed Countries** | 0.11 | 0 | 0.48 | 0.1 |

# Appendix II

# Supplementary material

## Compilation of data on urban–rural energy consumption/emission differentials from multiple sources

**China:** Pachauri, S. and Jiang, L. (2008). Interim Report IR-08-009, The Household Energy Transition in India and China, IIASA and Pan J., (ud). Rural Energy Patterns in China: A preliminary assessment from available data sources, Global Change and Economic Development Programme, The Chinese Academy of Social Sciences.

**India:** Pachauri, S. and Jiang, L. (2008). *The household energy transition in India and China: Interim report IR-08-009*. Vienna: IIASA.

**Thailand:** Choen, K. (2008). *Measuring rural–urban disparity toward promoting integrated development prospect between rural and urban areas in Thailand*. Retrieved from www.slideshare.net/choenkrainara/measuring-ruralurban-disparity-toward-promoting-integrated-development-prospects-between-rural-and-urban-areas-in-thailand on 5 April 2014.

**Canada:** Norman, J., MacLean, H.L. and Kennedy, C.A. (2006). Comparing high and low residential density: Lifecycle analysis of energy use and greenhouse gas emissions. *Journal of Urban Planning and Development, 132*(1), 10–21.

**United States:** Parshall, L., Gurney, K., Hammer, S.A., Mendoza, D., Zhou, Y. and Geethakumar, S. (2009). Modeling energy consumption and $CO_2$ emissions at the urban scale: Methodological challenges and insights from the United States. *Energy Policy, 38*(9), 4765–82. doi:10.1016/ j.enpol.2009.07.006.

**South American countries:** OLADE (2008). *Energy-Economic Information System (SIEE)*. Quito, Equador: Latin American Energy Organization. (Data only show electricity access.)

**European countries:** Meirmans, K. (2013). *Household direct energy consumption and $CO_2$ emissions in European countries*. Retrieved from http://ivem.eldoc.ub.rug.nl/FILES/ ivempubs/dvrapp/EES-2013/EES-2013-171T/EES-2013-171T_KoenMeirmans.pdf on 5 May 2014.

Heinonen, J. and Junnila, S. (2011). A carbon consumption comparison of rural and urban lifestyles. *Sustainability, 3*, 1234–49. doi:10.3390/su3081234.

Groenenberg, H., van Breevoort, P., Deng, Y., Noothout, P., van den Bos, A. and van Melle, T. (2011). *Rural energy in the EU: Country studies for France, Germany, Italy, Poland and the UK*. Retrieved from www.ecofys.com/files/files/ecofysreportruralenergyintheeu09_2011.pdf on 22 April 2014.

# Appendix III

## Global metadata of urbanization, carbon dioxide emissions, economy and energy use of 209 countries/territories

| | Urban population (percentage of total population) | | | | | | Carbon dioxide emissions (metric tons per capita) | | | | | | GDP per capita at PPP (2005 constant international dollars) | | | | | | Energy use (kg of oil equivalent per capita) 1971, 26 vs 109 | | | | | |
|---|---|---|---|---|---|---|---|---|---|---|---|---|---|---|---|---|---|---|---|---|---|---|---|---|
| | 1960 | 1970 | 1980 | 1990 | 2000 | 2010 | 1960 | 1970 | 1980 | 1990 | 2000 | 2010 | 1960 | 1970 | 1980 | 1990 | 2000 | 2010 | 1960 | 1970 | 1980 | 1990 | 2000 | 2010 |
| **LDC - LEAST DEVELOPED COUNTRIES** | | | | | | | | | | | | | | | | | | | | | | | | |
| Afghanistan | 8.0 | 11.0 | 15.7 | 18.2 | 20.6 | 23.2 | 0.0 | 0.2 | 0.1 | 0.2 | 0.0 | 0.3 | 61 | 159 | 276 | | | 561 | | | | | | |
| Angola | 10.4 | 15.0 | 24.3 | 37.1 | 49.0 | 58.4 | 0.1 | 0.6 | 0.7 | 0.4 | 0.7 | 1.6 | | | | 993 | 656 | 4219 | | 637 | 597 | 569 | 539 | 684 |
| Bangladesh | 5.1 | 7.6 | 14.9 | 19.8 | 23.6 | 27.9 | | | 0.1 | 0.1 | 0.2 | 0.4 | 86 | 136 | 220 | 284 | 356 | 664 | | 84 | 102 | 119 | 140 | 204 |
| Benin | 9.3 | 16.7 | 27.3 | 34.5 | 38.3 | 44.3 | 0.1 | | 0.1 | 0.1 | 0.2 | 0.5 | 93 | 115 | 378 | 392 | 339 | 690 | | 372 | 364 | 332 | 285 | 384 |
| Bhutan | 3.6 | 6.1 | 10.1 | 16.4 | 25.4 | 34.8 | | 0.0 | 0.1 | 0.2 | 0.7 | 0.7 | | | 329 | 560 | 778 | 2211 | | | | | 104 | |
| Burkina Faso | 4.7 | 5.7 | 8.8 | 13.8 | 17.8 | 25.7 | 0.0 | 0.0 | 0.0 | 0.1 | 0.1 | 0.1 | 68 | 81 | 283 | 352 | 225 | 593 | | | | | | |
| Burundi | 2.0 | 2.4 | 4.3 | 6.3 | 8.2 | 10.6 | | 0.0 | 0.0 | 0.0 | 0.0 | 0.0 | 70 | 70 | 223 | 202 | 130 | 220 | | | | | | |
| Cambodia | 10.3 | 16.0 | 9.0 | 15.5 | 18.6 | 19.8 | 0.0 | 0.2 | 0.0 | 0.0 | 0.2 | 0.3 | 111 | 102 | | | 299 | 783 | | | | | 279 | 350 |
| Central African Republic | 20.1 | 27.3 | 33.9 | 36.8 | 37.6 | 38.8 | 0.0 | 0.1 | 0.0 | 0.1 | 0.1 | 0.1 | 75 | 103 | 350 | 495 | 251 | 457 | | | | | | |
| Chad | 6.7 | 11.6 | 18.8 | 20.8 | 21.5 | 21.7 | 0.0 | 0.0 | 0.0 | 0.0 | 0.0 | 0.0 | 104 | 129 | 229 | 292 | 167 | 909 | | | | | | |
| Comoros | 12.6 | 19.4 | 23.2 | 27.9 | 28.1 | 28.0 | 0.1 | 0.1 | 0.2 | 0.2 | 0.2 | 0.2 | | | 394 | 606 | 382 | 795 | | | | | 43 | |
| Congo, Dem. Rep. | 22.3 | 30.3 | 28.7 | 27.7 | 29.3 | 33.7 | 0.2 | 0.1 | 0.1 | 0.1 | 0.0 | 0.0 | 220 | 244 | 546 | 268 | 92 | 211 | | 325 | 321 | 338 | 355 | 382 |
| Djibouti | 50.3 | 61.8 | 72.1 | 75.7 | 76.5 | 77.0 | 0.5 | 0.9 | 1.0 | 0.7 | 0.6 | 0.6 | | | | 767 | 763 | | | | | 220 | | |
| Equatorial Guinea | 25.5 | 27.0 | 27.9 | 34.7 | 38.8 | 39.3 | 0.1 | 0.1 | 0.3 | 0.3 | 0.9 | 6.7 | | 228 | | 353 | 2398 | 17613 | | | | | | |
| Eritrea | 9.8 | 12.6 | 14.4 | 15.8 | 17.6 | 20.9 | | | | 0.1 | 0.2 | 0.1 | | | | | 179 | 369 | | | | | 180 | 130 |
| Ethiopia | 6.4 | 8.6 | 10.4 | 12.6 | 14.7 | 16.8 | 0.0 | 0.1 | 0.1 | 0.1 | 0.1 | 0.1 | | | | 251 | 123 | 302 | | 414 | 409 | 411 | 382 | |
| Gambia, The | 12.1 | 19.5 | 28.4 | 38.3 | 48.8 | 56.7 | 0.0 | 0.1 | 0.3 | 0.2 | 0.2 | 0.3 | | 117 | 399 | 346 | 637 | 566 | | | | | 67 | |
| Guinea | 10.5 | 16.0 | 23.6 | 28.0 | 31.0 | 35.0 | 0.1 | 0.2 | 0.2 | 0.2 | 0.1 | 0.1 | | | | 443 | 342 | 435 | | | | | | |
| Guinea-Bissau | 13.6 | 15.1 | 17.6 | 28.1 | 35.9 | 43.2 | 0.0 | 0.1 | 0.2 | 0.2 | 0.1 | 0.2 | | 115 | 135 | 240 | 284 | 527 | | | | | 73 | |
| Haiti | 15.6 | 19.8 | 20.5 | 28.5 | 35.6 | 52.0 | 0.1 | 0.1 | 0.1 | 0.1 | 0.2 | 0.2 | | 326 | 512 | 400 | 427 | 670 | | 314 | 366 | 219 | 234 | 243 |
| Kiribati | 16.3 | 24.1 | 32.3 | 35.0 | 43.0 | 43.8 | | 0.5 | 0.5 | 0.3 | 0.4 | 0.6 | | | | | 815 | 1539 | | | | | 100 | |
| Lao PDR | 7.9 | 9.6 | 12.4 | 15.4 | 22.0 | 33.1 | 0.0 | 0.2 | 0.1 | 0.1 | 0.2 | 0.3 | | | | 204 | 321 | 1123 | | | | | | |
| Lesotho | 3.4 | 8.6 | 11.5 | 14.0 | 20.0 | 26.8 | | | | | | 0.0 | 41 | 67 | 330 | 341 | 415 | 1097 | | | | | | |
| Liberia | 18.6 | 26.0 | 35.2 | 40.9 | 44.3 | 47.8 | 0.1 | 1.0 | 1.1 | 0.2 | 0.2 | 0.0 | 170 | 228 | 452 | 183 | 183 | 327 | | | | | | |
| Madagascar | 10.6 | 14.1 | 18.5 | 23.6 | 27.1 | 31.9 | 0.1 | 0.1 | 0.2 | 0.1 | 0.1 | 0.1 | 132 | 169 | 462 | 267 | 246 | 419 | | | | | | |
| Malawi | 4.4 | 6.1 | 9.1 | 11.6 | 14.6 | 15.5 | | 0.1 | 0.1 | 0.1 | 0.1 | 0.1 | 46 | 64 | 198 | 199 | 154 | 360 | | | | | | |
| Maldives | 11.2 | 11.9 | 22.3 | 25.8 | 27.7 | 40.0 | | | 0.3 | 0.7 | 1.8 | 3.3 | | | 275 | 996 | 2289 | 6552 | | | | | 236 | |
| Mali | 11.1 | 14.3 | 18.5 | 23.3 | 28.1 | 34.3 | 0.0 | 0.0 | 0.1 | 0.1 | 0.1 | 0.0 | | 63 | 265 | 304 | 236 | 674 | | | | | | |

| | | | | | | | | | | | | | | | | | | | | | | | |
|---|---|---|---|---|---|---|---|---|---|---|---|---|---|---|---|---|---|---|---|---|---|---|---|
| Mauritania | 6.9 | 14.6 | 27.4 | 39.7 | 40.0 | 41.2 | 0.0 | 0.4 | 0.4 | 1.3 | 0.5 | 0.6 | 108 | 182 | 462 | 504 | 478 | 1017 | 713 | 553 | 436 | 392 | 412 |
| Mozambique | 3.7 | 5.8 | 13.1 | 21.1 | 29.1 | 31.0 | 0.2 | 0.3 | 0.3 | 0.1 | 0.1 | 0.1 | | | 290 | 185 | 236 | 387 | 284 | 273 | 254 | 265 | 270 |
| Myanmar | 19.2 | 22.8 | 24.0 | 24.6 | 27.2 | 32.1 | 0.1 | 0.2 | 0.1 | 0.1 | 0.2 | 0.2 | | | | | | | 310 | 317 | 320 | 350 | 381 |
| Nepal | 3.5 | 4.0 | 6.1 | 8.9 | 13.4 | 16.7 | 0.0 | 0.0 | 0.0 | 0.1 | 0.1 | 0.1 | 53 | 75 | 135 | 200 | 237 | 596 | | | | | |
| Niger | 5.8 | 8.8 | 13.4 | 15.4 | 16.2 | 17.6 | 0.0 | 0.0 | 0.1 | 0.1 | 0.1 | 0.1 | 135 | 147 | 430 | 320 | 164 | 360 | | | | | |
| Rwanda | 2.4 | 3.2 | 4.7 | 5.4 | 13.8 | 18.8 | 0.0 | 0.0 | 0.1 | 0.1 | 0.1 | 0.1 | 41 | 59 | 244 | 353 | 207 | 519 | | | | | |
| Samoa | 18.9 | 20.4 | 21.2 | 21.2 | 22.0 | 20.1 | 0.1 | 0.2 | 0.6 | 0.8 | 0.8 | 0.9 | | | | 688 | 1373 | 3076 | | | 265 | 200 | |
| Sao Tome and Principe | 16.1 | 29.5 | 33.5 | 43.6 | 53.4 | 62.0 | 0.2 | 0.2 | 0.4 | 0.4 | 0.3 | 0.6 | | | | | 550 | 1128 | | | | | 265 |
| Senegal | 23.0 | 30.0 | 35.8 | 38.9 | 40.3 | 42.3 | 0.3 | 0.3 | 0.6 | 0.4 | 0.4 | 0.5 | 249 | 243 | 629 | 761 | 475 | 999 | 286 | 280 | 224 | 243 | 265 |
| Sierra Leone | 17.4 | 23.4 | 29.1 | 33.0 | 35.8 | 38.9 | 0.3 | 0.3 | 0.2 | 0.1 | 0.1 | 0.1 | 149 | 172 | 346 | 161 | 154 | 448 | | | | | |
| Solomon Islands | 5.8 | 8.9 | 10.6 | 13.7 | 15.8 | 20.0 | 0.1 | 0.3 | 0.3 | 0.5 | 0.4 | 0.4 | 65 | 92 | 732 | 970 | 1055 | 1289 | | | 171 | | |
| Somalia | 17.3 | 22.7 | 26.8 | 29.7 | 33.2 | 37.3 | 0.0 | 0.1 | 0.1 | 0.0 | 0.1 | 0.1 | | | | | 99 | 145 | | | | | |
| South Sudan | 8.7 | 8.6 | 8.5 | 13.3 | 16.5 | 17.9 | | | | | | | | | | | | 1527 | | | | | |
| Sudan | 10.7 | 16.5 | 20.0 | 28.6 | 32.5 | 33.1 | 0.1 | 0.3 | 0.2 | 0.2 | 0.2 | 0.3 | 107 | 151 | 398 | 481 | 356 | 1422 | 491 | 438 | 412 | 388 | 364 |
| Tanzania | 5.2 | 7.9 | 14.6 | 18.9 | 24.3 | 26.3 | 0.1 | 0.1 | 0.1 | 0.1 | 0.1 | 0.2 | | | | 172 | 308 | 525 | 539 | 429 | 382 | 394 | 446 |
| Timor-Leste | 10.1 | 12.9 | 16.5 | 20.8 | 24.3 | 28.0 | | | | | | 0.2 | 77 | 120 | 418 | 430 | 266 | 503 | | | | | |
| Togo | 10.1 | 21.3 | 24.7 | 28.6 | 32.9 | 37.5 | 0.0 | 0.1 | 0.3 | 0.2 | 0.3 | 0.2 | | | | | | | 333 | 326 | 334 | 434 | 427 |
| Tuvalu | 15.9 | 22.1 | 29.8 | 40.7 | 46.0 | 50.1 | | | | | | | | | | 980 | 1459 | 3238 | | | | | |
| Uganda | 4.4 | 6.7 | 7.5 | 11.1 | 12.1 | 15.2 | 0.1 | 0.2 | 0.0 | 0.0 | 0.1 | 0.1 | 62 | 99 | 245 | 255 | | 472 | | | | | |
| Vanuatu | 10.4 | 12.3 | 14.7 | 18.7 | 21.7 | 24.6 | 0.0 | 0.5 | 0.5 | 0.5 | 0.4 | 0.8 | | | 981 | 1080 | 1470 | 2966 | | | 159 | | |
| Yemen, Rep. | 9.1 | 13.3 | 16.5 | 20.9 | 26.3 | 31.7 | 0.4 | 0.4 | 0.8 | 0.8 | 0.8 | 1.0 | | | | 479 | 550 | 1401 | 119 | 161 | 213 | 271 | 367 |
| Zambia | 18.1 | 30.4 | 39.8 | 39.4 | 34.8 | 38.7 | 0.7 | 0.9 | 0.6 | 0.3 | 0.2 | 0.2 | 227 | 427 | 664 | 419 | 322 | 1225 | 808 | 769 | 688 | 618 | 609 |
| **MDR - DEVELOPED COUNTRIES** | | | | | | | | | | | | | | | | | | | | | | | |
| Albania | 30.7 | 31.7 | 33.8 | 36.4 | 41.7 | 52.3 | 1.3 | 1.8 | 1.9 | 2.2 | 0.9 | 1.4 | | | | 610 | 1115 | 3764 | 783 | 1123 | 775 | 534 | 654 |
| Andorra | 58.5 | 80.2 | 92.1 | 94.7 | 92.4 | 87.8 | 6.6 | 8.0 | 8.0 | | | | 3238 | | | 18877 | 17334 | | | | | | |
| Australia | 81.5 | 85.3 | 85.8 | 85.4 | 87.2 | 89.0 | 8.6 | 11.8 | 15.0 | 16.8 | 17.2 | 16.9 | 1811 | 3305 | 10207 | 18247 | 21678 | 51825 | 3990 | 4737 | 5053 | 5645 | 5561 |
| Austria | 64.7 | 65.3 | 65.4 | 65.8 | 65.8 | 67.5 | 4.4 | 6.8 | 6.9 | 7.9 | 8.0 | 8.0 | 935 | 2038 | 10758 | 21458 | 23974 | 44723 | 2509 | 3067 | 3236 | 3565 | 4080 |
| Belarus | 32.4 | 44.0 | 56.5 | 66.0 | 70.0 | 74.6 | | | | | 5.3 | 6.6 | | | | 1705 | 1273 | 5819 | | | 4465 | 2468 | 2917 |
| Belgium | 92.5 | 93.8 | 95.4 | 96.4 | 97.1 | 97.5 | 9.9 | 13.0 | 13.7 | 10.9 | 11.3 | 10.0 | 1274 | 2732 | 12707 | 20350 | 22697 | 42960 | 4100 | 4744 | 4844 | 5707 | 5589 |
| Bermuda | 100.0 | 100.0 | 100.0 | 100.0 | 100.0 | 100.0 | 3.6 | 4.1 | 8.0 | 10.1 | 8.0 | 7.3 | 1902 | 3387 | 11218 | 26842 | 56284 | 88207 | | | | | |
| Bosnia and Herzegovina | 19.0 | 27.2 | 35.5 | 39.2 | 43.0 | 47.7 | | | | | 6.1 | 8.1 | | | | | 1436 | 4362 | | | 1550 | 1133 | 1677 |
| Bulgaria | 37.1 | 52.3 | 62.1 | 66.4 | 68.9 | 72.5 | 2.8 | 7.2 | 8.7 | 8.7 | 5.3 | 6.0 | | 2261 | 2377 | 1579 | | 6453 | 2229 | 3204 | 3236 | 2287 | 2420 |
| Canada | 69.1 | 75.7 | 75.7 | 76.6 | 79.5 | 80.6 | 10.8 | 16.0 | 17.4 | 16.2 | 17.4 | 14.7 | 2295 | 4047 | 10934 | 20968 | 23560 | 46376 | 6530 | 7805 | 7505 | 8172 | 7381 |
| Channel Islands | 38.7 | 36.1 | 32.2 | 31.4 | 30.5 | 31.1 | | | | | 4.4 | 4.7 | | | 5185 | 4862 | 43299 | 13327 | | | 1888 | 1760 | 1938 |
| Croatia | 30.2 | 40.2 | 50.1 | 54.0 | 55.6 | 57.5 | | | | | | | | | | | | | | | | | |
| Curacao | | | | | | | | | | | | | | | | | | | | | | | |
| Czech Republic | 59.5 | 64.4 | 75.2 | 74.0 | 75.2 | 73.5 | 6.5 | 12.6 | 11.8 | 9.7 | 12.2 | 10.7 | 1365 | 3366 | 13607 | 26423 | 29980 | 56452 | 4618 | 4556 | 4797 | 3997 | 4205 |
| Denmark | 73.7 | 79.7 | 83.7 | 84.8 | 85.1 | 86.8 | | | | | 8.9 | 8.3 | 3787 | | | | | | 3729 | 3735 | 3377 | 3490 | 3480 |
| Estonia | 57.5 | 64.9 | 69.7 | 71.1 | 69.3 | 69.5 | | | | | 11.0 | 13.7 | | | | | 4114 | 22850 | | | 6316 | 3418 | 4165 |
| Faeroe Islands | 21.4 | 28.0 | 31.2 | 30.6 | 36.3 | 40.9 | 1.7 | 6.7 | 9.8 | 13.0 | 15.3 | 14.3 | | | | | | | | | | | |
| Finland | 55.3 | 63.7 | 71.7 | 79.4 | 82.2 | 83.6 | 3.4 | 8.8 | 12.2 | 10.4 | 10.1 | 11.5 | 1179 | 2436 | 11091 | 27852 | 23530 | 43846 | 3939 | 5147 | 5692 | 6227 | 6792 |

*(continued)*

*(continued)*

| | | | | | | | | | | | | | | | | | | | | | | | |
|---|---|---|---|---|---|---|---|---|---|---|---|---|---|---|---|---|---|---|---|---|---|---|---|
| France | 61.9 | 71.1 | 73.3 | 74.1 | 76.9 | 85.2 | 5.8 | 8.5 | 9.2 | 6.8 | 6.0 | 5.6 | 1343 | 2822 | 12500 | 21301 | 21775 | 39186 | 3031 | 3473 | 3835 | 4137 | 4016 |
| Germany | 71.4 | 72.3 | 72.8 | 73.1 | 73.1 | 73.8 | | | | | 10.1 | 9.1 | | 2672 | 11746 | 21584 | 22946 | 40145 | 3895 | 4562 | 4421 | 4094 | 4033 |
| Greece | 42.9 | 52.5 | 57.7 | 58.8 | 59.7 | 61.2 | 2.7 | 8.2 | 5.3 | 7.2 | 8.4 | 7.7 | 534 | 1425 | 5620 | 9190 | 11396 | 25851 | 984 | 1554 | 2111 | 2481 | 2442 |
| Greenland | 58.5 | 72.7 | 76.1 | 79.7 | 81.6 | 84.4 | 6.9 | 8.2 | 11.2 | 9.5 | 11.1 | | | 1498 | 9483 | 18327 | 19004 | | | | | | |
| Hungary | 55.9 | 60.1 | 64.2 | 65.8 | 64.6 | 69.0 | 1.1 | 4.5 | 8.1 | 6.1 | 5.6 | 6.2 | 1415 | 536 | 2069 | 3186 | 4543 | 12750 | 1836 | 2647 | 2772 | 2448 | 2567 |
| Iceland | 80.3 | 84.9 | 88.3 | 90.8 | 92.4 | 93.6 | 6.9 | 6.8 | 8.2 | 7.8 | 7.7 | 6.2 | | 2538 | 14602 | 25009 | 30929 | 39507 | 4378 | 6561 | 8196 | 11023 | 16882 |
| Ireland | 45.8 | 51.7 | 55.3 | 56.9 | 59.1 | 61.9 | 4.0 | 6.6 | 7.7 | 8.9 | 10.9 | 8.8 | 686 | 1460 | 6258 | 13779 | 25579 | 45617 | 2245 | 2413 | 2809 | 3567 | 3118 |
| Isle of Man | 55.1 | 55.8 | 51.8 | 51.7 | 51.8 | 50.6 | | | | | | | | | | | 20359 | | | | | | |
| Italy | 59.4 | 64.3 | 66.6 | 66.7 | 67.2 | 68.2 | 2.2 | 5.5 | 6.9 | 7.4 | 7.9 | 6.7 | 804 | 2030 | 8148 | 20065 | 19388 | 33761 | 1949 | 2318 | 2584 | 3012 | 2815 |
| Japan | 63.3 | 71.9 | 76.2 | 77.3 | 78.6 | 90.5 | 2.5 | 7.4 | 8.1 | 8.9 | 9.6 | 9.2 | 479 | 2004 | 9308 | 25124 | 37292 | 43118 | 2531 | 2950 | 3556 | 4091 | 3916 |
| Latvia | 52.9 | 60.7 | 67.1 | 69.3 | 68.1 | 67.7 | | | | | 2.6 | 3.6 | | | | 2796 | 3309 | 11447 | | | 2949 | 1618 | 2214 |
| Liechtenstein | 20.4 | 18.5 | 18.3 | 16.9 | 15.1 | 14.4 | | | | | | | | | | | 3267 | 2841 | | | | | |
| Lithuania | 39.5 | 49.6 | 61.2 | 67.6 | 67.0 | 67.0 | | | | | 3.5 | 4.4 | | 4237 | 20669 | 49452 | 75058 | 11722 | | | 4344 | 2038 | 2277 |
| Luxembourg | 69.6 | 74.4 | 80.0 | 80.9 | 83.8 | 85.2 | 36.7 | 40.5 | 30.3 | 26.2 | 18.9 | 21.4 | 2242 | 4261 | 16389 | 33177 | 46453 | 102009 | 11863 | 9775 | 8874 | 7644 | 8324 |
| Macedonia, FYR | 34.0 | 47.1 | 53.5 | 57.8 | 59.4 | 59.2 | | | | 5.2 | 5.9 | 5.2 | | | | 2225 | 1748 | 4442 | | | 1233 | 1300 | 1371 |
| Malta | 90.1 | 89.7 | 89.8 | 90.4 | 92.4 | 94.7 | 1.0 | 2.2 | 3.2 | 6.2 | 5.4 | 6.2 | | 828 | 3948 | 7192 | 10377 | 19695 | 696 | 1005 | 1963 | 1773 | 2046 |
| Moldova | 23.4 | 32.1 | 40.4 | 46.8 | 44.6 | 46.9 | | | | | 1.0 | 1.4 | | | 972 | 354 | 1632 | | | | 2677 | 792 | 962 |
| Monaco | 100.0 | 100.0 | 100.0 | 100.0 | 100.0 | 100.0 | | | | | | | | 12480 | 51527 | 84290 | 82537 | 145230 | | | | | |
| Montenegro | 18.8 | 26.9 | 36.8 | 48.0 | 58.5 | 63.1 | | | | | | 4.2 | | | | | 1610 | 6636 | | | | | 1893 |
| Netherlands | 59.8 | 61.7 | 64.7 | 68.7 | 76.8 | 82.7 | 6.4 | 10.9 | 12.5 | 10.9 | 10.4 | 11.0 | 1069 | 2711 | 12775 | 19722 | 24180 | 46468 | 3855 | 4549 | 4393 | 4598 | 5021 |
| New Zealand | 76.0 | 81.1 | 83.4 | 84.7 | 85.7 | 86.2 | 4.9 | 5.1 | 5.6 | 7.1 | 8.5 | 7.2 | 2313 | 2302 | 7401 | 13527 | 13475 | 32796 | 2423 | 2886 | 3865 | 4421 | 4187 |
| Norway | 49.9 | 65.4 | 70.5 | 72.0 | 76.1 | 79.1 | 3.7 | 7.2 | 8.6 | 7.4 | 8.6 | 11.7 | 1442 | 3284 | 15595 | 27732 | 37473 | 86156 | 3407 | 4483 | 4952 | 5810 | 6614 |
| Poland | 47.9 | 52.1 | 58.1 | 61.3 | 61.7 | 60.9 | 6.7 | 9.3 | 13.1 | 9.6 | 7.8 | 8.3 | | 1694 | 4454 | 12302 | | | 2627 | 3559 | 2705 | 2317 | 2659 |
| Portugal | 35.0 | 38.8 | 42.8 | 47.9 | 54.4 | 60.5 | 0.9 | 1.8 | 2.8 | 4.2 | 6.1 | 5.0 | 360 | 922 | 3323 | 7779 | 11399 | 21512 | 726 | 1022 | 1677 | 2398 | 2226 |
| Romania | 34.2 | 40.3 | 46.1 | 53.2 | 53.0 | 52.8 | 2.9 | 5.9 | 8.8 | 6.8 | 4.0 | 3.9 | | | 1651 | 1662 | 8139 | | 2059 | 2933 | 2683 | 1614 | 1730 |
| Russian Federation | 53.7 | 62.5 | 69.8 | 73.4 | 73.4 | 73.7 | | | | | 10.6 | 12.2 | | | | 3485 | 1775 | 10710 | | | 5929 | 4233 | 4932 |
| San Marino | 48.9 | 59.8 | 81.2 | 90.4 | 93.4 | 94.1 | | | | | | | | | | | | 28696 | | | | | |
| Serbia | | | | | 53.0 | 56.0 | | | | 6.3 | 6.7 | 7.5 | | | | | 809 | 5073 | | | 2599 | 1826 | 2131 |
| Slovak Republic | 33.5 | 41.1 | 51.6 | 56.5 | 56.2 | 54.8 | | | | | | | | | | 2211 | 5330 | 16151 | 3129 | 3986 | 4025 | 3293 | 3307 |
| Slovenia | 28.2 | 37.0 | 48.0 | 50.4 | 50.8 | 50.0 | | | | | | | | | | 8699 | 10045 | 22898 | | | 2858 | 3224 | 3529 |
| Spain | 56.6 | 66.0 | 72.8 | 75.4 | 76.3 | 77.3 | 1.6 | 3.5 | 5.7 | 5.6 | 7.3 | 5.8 | 396 | 1176 | 6031 | 13397 | 14414 | 29539 | 1246 | 1808 | 2319 | 3026 | 2743 |
| Sweden | 72.5 | 81.0 | 83.1 | 83.1 | 84.0 | 85.1 | 6.6 | 11.5 | 8.6 | 6.0 | 5.6 | 5.0 | 1983 | 4493 | 16221 | 29026 | 27865 | 49360 | 4450 | 4872 | 5515 | 5360 | 5472 |
| Switzerland | 51.0 | 57.4 | 57.1 | 53.2 | 73.3 | 73.6 | 3.7 | 6.5 | 6.4 | 6.4 | 5.4 | 6.6 | 1787 | 3714 | 17807 | 36337 | 35639 | 70370 | 2638 | 3171 | 3628 | 3480 | 3348 |
| Ukraine | 46.8 | 54.8 | 61.7 | 66.8 | 67.1 | 68.7 | 11.2 | 11.7 | 10.3 | 10.0 | 9.2 | 7.9 | | | | 1570 | 636 | 2974 | | | 4856 | 2721 | 2884 |
| United Kingdom | 78.4 | 77.1 | 78.5 | 78.1 | 78.7 | 79.5 | 16.0 | 21.1 | 20.8 | 19.1 | 20.2 | 17.6 | 1380 | 2242 | 9623 | 17805 | 25362 | 36425 | 3733 | 3524 | 3597 | 3786 | 3217 |
| United States | 70.0 | 73.6 | 73.7 | 75.3 | 79.1 | 82.1 | | | | | | | 2881 | 5247 | 12598 | 23955 | 36467 | 48358 | 7645 | 7942 | 7672 | 8057 | 7162 |
| **LDR MINUS LDC** | | | | | | | | | | | | | | | | | | | | | | | |
| Algeria | 30.5 | 39.5 | 43.5 | 52.1 | 60.8 | 72.0 | 0.5 | 1.0 | 3.4 | 3.0 | 2.8 | 3.3 | 242 | 331 | 2174 | 2365 | 1727 | 4350 | 229 | 575 | 846 | 851 | 1082 |
| American Samoa | 66.2 | 70.4 | 74.3 | 80.9 | 88.8 | 93.0 | | | | | | | | | | | | | | | | | |
| Antigua and Barbuda | 39.7 | 33.8 | 34.6 | 35.4 | 32.1 | 29.9 | 0.7 | 7.1 | 2.0 | 4.9 | 4.4 | 5.9 | | 1566 | 6325 | 10144 | 13315 | | | 1596 | | | |

| | | | | | | | | | | | | | | | | | | | | | | | |
|---|---|---|---|---|---|---|---|---|---|---|---|---|---|---|---|---|---|---|---|---|---|---|---|
| Argentina | 73.6 | 78.9 | 82.9 | 87.0 | 90.1 | 92.3 | 2.4 | 3.5 | 3.9 | 3.5 | 3.8 | 4.5 | | 1317 | 2737 | 4333 | 7701 | 9133 | 1381 | 1487 | 1412 | 1652 | 1936 |
| Armenia | 51.3 | 59.9 | 66.1 | 67.4 | 64.7 | 64.1 | | | | | 1.1 | 1.4 | | | | 637 | 621 | 3125 | | | 2175 | 655 | 838 |
| Aruba | 50.8 | 50.6 | 50.5 | 50.3 | 46.7 | 46.8 | | | | 29.6 | 24.6 | 22.8 | | | | | 20620 | 24289 | | | | | |
| Azerbaijan | 52.7 | 50.0 | 52.8 | 53.7 | 51.4 | 53.4 | | | | | 3.7 | 5.1 | | | | 1237 | 655 | 5843 | | | 3165 | 1403 | 1280 |
| Bahamas, The | 59.7 | 66.7 | 73.1 | 79.8 | 82.0 | 84.1 | 3.7 | 15.2 | 37.9 | 7.6 | 5.6 | 6.8 | 1550 | 3179 | 6339 | 12351 | 21251 | 21881 | | | 2517 | | |
| Bahrain | 82.3 | 83.8 | 86.1 | 88.1 | 88.4 | 88.6 | 3.5 | 12.2 | 21.9 | 24.0 | 27.9 | 19.3 | | | 8538 | 8529 | 13563 | 20546 | 6380 | 7794 | 8771 | 8776 | 7568 |
| Barbados | 36.8 | 37.5 | 39.6 | 32.7 | 38.3 | 43.9 | 0.7 | 1.8 | 2.7 | 4.1 | 4.4 | 5.4 | 379 | 761 | 3459 | 7802 | 11675 | 15812 | | | 1242 | | |
| Belize | 54.0 | 51.0 | 49.4 | 47.5 | 47.7 | 45.0 | 0.5 | 1.0 | 1.3 | 1.7 | 2.9 | 1.4 | 305 | 436 | 1351 | 2202 | 3486 | 4530 | | | 566 | | |
| Bolivia | 36.8 | 39.8 | 45.5 | 55.6 | 61.8 | 66.4 | 0.3 | 0.6 | 0.9 | 0.8 | 1.2 | 1.5 | 168 | 241 | 845 | 716 | 989 | 1935 | 236 | 455 | 384 | 440 | 723 |
| Botswana | 3.1 | 7.8 | 16.5 | 41.9 | 53.2 | 61.0 | | | 1.0 | 1.6 | 2.4 | 2.7 | 58 | 139 | 1064 | 2739 | 3297 | 6980 | | | 911 | 1046 | 1149 |
| Brazil | 46.1 | 55.9 | 65.5 | 73.9 | 81.2 | 84.3 | 0.6 | 1.0 | 1.5 | 1.4 | 1.9 | 2.2 | 208 | 441 | 1931 | 3087 | 3694 | 10978 | 709 | 935 | 937 | 1074 | 1362 |
| Brunei Darussalam | 43.4 | 61.7 | 59.9 | 65.8 | 71.2 | 75.6 | 4.1 | 63.3 | 35.7 | 25.0 | 19.7 | 22.9 | | 1381 | 25531 | 13702 | 18087 | 30880 | 1306 | 6991 | 6721 | 7187 | 8089 |
| Cabo Verde | 16.7 | 19.6 | 23.5 | 44.1 | 53.4 | 61.8 | 0.1 | 0.1 | 0.4 | 0.3 | 0.4 | 0.7 | | | 472 | 1021 | 1367 | 3413 | | | 83 | | |
| Cameroon | 13.9 | 20.3 | 31.9 | 39.7 | 45.5 | 51.5 | 0.1 | 0.1 | 0.4 | 0.1 | 0.2 | 0.4 | 115 | 171 | 755 | 924 | 583 | 1091 | 388 | 409 | 413 | 396 | 337 |
| Cayman Islands | 100.0 | 100.0 | 100.0 | 100.0 | 100.0 | 100.0 | 1.4 | 4.0 | 10.2 | 10.1 | 10.9 | 10.6 | | | | | | | | | | | |
| Chile | 67.8 | 75.2 | 81.2 | 83.3 | 85.9 | 88.9 | 1.8 | 2.6 | 2.3 | 2.6 | 3.8 | 4.2 | 551 | 938 | 2463 | 2388 | 5133 | 12685 | 892 | 847 | 1060 | 1629 | 1803 |
| China | 16.2 | 17.4 | 19.4 | 26.4 | 35.9 | 49.2 | 1.2 | 0.9 | 1.5 | 2.2 | 2.7 | 6.2 | 92 | 112 | 193 | 314 | 949 | 4433 | 466 | 610 | 767 | 920 | 1881 |
| Colombia | 45.0 | 54.8 | 62.1 | 68.3 | 72.1 | 75.0 | 1.0 | 1.3 | 1.6 | 1.7 | 1.5 | 1.6 | 252 | 337 | 1240 | 1209 | 2504 | 6179 | 633 | 657 | 727 | 647 | 694 |
| Congo, Rep. | 31.6 | 39.1 | 47.9 | 54.3 | 58.7 | 63.2 | 0.2 | 0.4 | 0.2 | 0.5 | 0.3 | 0.5 | 130 | 206 | 950 | 1174 | 1030 | 2920 | 370 | 345 | 325 | 260 | 368 |
| Costa Rica | 34.3 | 38.8 | 43.1 | 50.7 | 59.0 | 64.2 | 0.4 | 0.7 | 1.0 | 1.0 | 1.4 | 1.7 | 380 | 540 | 2057 | 2405 | 4058 | 7773 | 432 | 534 | 545 | 731 | 995 |
| Cote d'Ivoire | 17.7 | 28.2 | 36.8 | 39.3 | 43.5 | 50.6 | 0.1 | 0.5 | 0.8 | 0.5 | 0.4 | 0.3 | 157 | 278 | 1231 | 891 | 646 | 1208 | 447 | 432 | 357 | 417 | 516 |
| Cuba | 58.4 | 60.3 | 68.1 | 73.4 | 75.6 | 75.2 | 1.9 | 2.1 | 3.2 | 3.1 | 2.3 | 3.4 | | 653 | 2025 | 2702 | 2744 | 5702 | 1211 | 1523 | 1669 | 1154 | 1002 |
| Cyprus | 35.6 | 40.8 | 58.6 | 66.8 | 68.6 | 70.3 | 1.5 | 2.8 | 4.7 | 6.1 | 7.3 | 7.0 | | | 4232 | 9642 | 13422 | 27889 | 946 | 1262 | 1781 | 2265 | 2213 |
| Dominica | 27.3 | 46.9 | 63.4 | 67.7 | 67.2 | 67.1 | 0.2 | 0.4 | 0.5 | 0.8 | 1.5 | 1.9 | | | 785 | 2345 | 4657 | 6630 | | | 298 | | |
| Dominican Republic | 30.2 | 40.2 | 51.3 | 55.2 | 61.7 | 69.1 | 0.3 | 0.7 | 1.1 | 1.3 | 2.3 | 2.1 | 203 | 328 | 1138 | 976 | 2770 | 5166 | 503 | 590 | 563 | 866 | 717 |
| Ecuador | 33.9 | 39.3 | 47.0 | 55.1 | 60.3 | 66.9 | 0.4 | 0.7 | 1.7 | 1.7 | 1.7 | 2.2 | 224 | 475 | 2261 | 1505 | 1462 | 4501 | 361 | 632 | 577 | 624 | 828 |
| Egypt, Arab Rep. | 37.9 | 42.2 | 43.9 | 43.5 | 42.8 | 43.4 | 0.6 | 0.6 | 1.0 | 1.3 | 2.1 | 2.6 | 148 | 211 | 510 | 766 | 1510 | 2804 | 209 | 337 | 574 | 615 | 942 |
| El Salvador | 38.3 | 39.4 | 44.1 | 49.2 | 58.9 | 64.3 | 0.2 | 0.4 | 0.5 | 0.5 | 1.0 | 1.0 | | 303 | 767 | 898 | 2204 | 3444 | 457 | 540 | 462 | 666 | 677 |
| Fiji | 29.7 | 34.8 | 37.8 | 41.6 | 47.9 | 51.8 | 0.5 | 1.0 | 1.3 | 1.1 | 1.1 | 1.5 | 286 | 422 | 1893 | 1836 | 2075 | 3748 | | | 392 | | |
| French Polynesia | 42.4 | 55.3 | 57.4 | 55.9 | 52.4 | 51.4 | 0.5 | 1.8 | 1.9 | 3.2 | 2.7 | 3.3 | | 2299 | 8979 | 16037 | 14530 | | | | | | |
| Gabon | 17.4 | 32.0 | 54.7 | 69.1 | 80.1 | 85.8 | 0.3 | 3.5 | 9.1 | 5.1 | 0.9 | 1.7 | 284 | 550 | 5891 | 6287 | 4135 | 9322 | 1782 | 1891 | 1248 | 1194 | 1275 |
| Georgia | 43.1 | 48.0 | 52.5 | 55.0 | 52.6 | 52.7 | | | | | 1.0 | 1.4 | | | | 1611 | 692 | 2614 | | | 2586 | 649 | 701 |
| Ghana | 23.3 | 29.0 | 31.2 | 36.4 | 44.0 | 51.2 | 0.2 | 0.3 | 0.2 | 0.3 | 0.3 | 0.4 | 183 | 258 | 412 | 403 | 265 | 1326 | 339 | 372 | 362 | 411 | 413 |
| Grenada | 30.3 | 32.2 | 32.9 | 33.4 | 35.9 | 38.8 | 0.2 | 0.5 | 0.5 | 1.1 | 1.9 | 2.5 | | | 939 | 2296 | 5149 | 7486 | | | 441 | | |
| Guam | 50.1 | 61.9 | 93.8 | 90.8 | 93.1 | 93.2 | | | | | | | | | | | | | | | | | |
| Guatemala | 31.1 | 35.5 | 37.4 | 41.1 | 45.1 | 49.3 | 0.3 | 0.4 | 0.6 | 0.6 | 0.9 | 0.8 | 252 | 352 | 1125 | 860 | 1722 | 2882 | 491 | 542 | 496 | 628 | 715 |
| Guyana | 29.0 | 29.4 | 30.5 | 29.6 | 28.7 | 28.3 | 1.2 | 2.2 | 2.3 | 1.6 | 2.2 | 2.2 | 304 | 372 | 776 | 547 | 957 | 2874 | | | 527 | | |
| Honduras | 22.7 | 28.9 | 34.9 | 40.5 | 45.5 | 51.6 | 0.3 | 0.5 | 0.6 | 0.5 | 0.8 | 1.1 | 168 | 269 | 706 | 622 | 1140 | 2064 | 502 | 514 | 485 | 479 | 599 |
| Hong Kong SAR, China | 85.2 | 87.7 | 91.5 | 99.5 | 100.0 | 100.0 | 1.0 | 2.1 | 3.3 | 4.8 | 6.1 | 5.2 | 429 | 960 | 5700 | 13486 | 25757 | 32558 | 742 | 914 | 1518 | 2009 | 1970 |
| India | 17.9 | 19.8 | 23.1 | 25.5 | 27.7 | 30.9 | 0.3 | 0.4 | 0.5 | 0.8 | 1.1 | 1.7 | 84 | 114 | 271 | 376 | 457 | 1417 | 276 | 294 | 365 | 439 | 600 |
| Indonesia | 14.6 | 17.1 | 22.1 | 30.6 | 42.0 | 49.9 | 0.2 | 0.3 | 0.7 | 0.8 | 1.3 | 1.8 | | 85 | 536 | 641 | 790 | 2947 | 300 | 383 | 552 | 741 | 878 |
| Iran, Islamic Rep. | 33.7 | 41.2 | 49.7 | 56.3 | 64.0 | 68.9 | 1.7 | 3.2 | 3.1 | 3.7 | 5.7 | 7.7 | | 370 | 2315 | 2059 | 1537 | 5675 | 565 | 979 | 1230 | 1866 | 2829 |
| Iraq | 42.9 | 56.2 | 65.5 | 69.7 | 67.8 | 66.5 | 1.1 | 2.4 | 3.3 | 3.0 | 3.0 | 3.7 | 231 | | | | | 4613 | 403 | 707 | 1125 | 1090 | 1222 |

*(continued)*

*(continued)*

| | C1 | C2 | C3 | C4 | C5 | C6 | C7 | C8 | C9 | C10 | C11 | C12 | C13 | C14 | C15 | C16 | C17 | C18 | C19 | C20 | C21 | C22 | C23 |
|---|---|---|---|---|---|---|---|---|---|---|---|---|---|---|---|---|---|---|---|---|---|---|---|
| Israel | 76.8 | 84.2 | 88.6 | 90.4 | 91.2 | 91.8 | 3.1 | 5.6 | 5.5 | 7.2 | 10.0 | 9.3 | 1366 | 1806 | 5617 | 11264 | 19859 | 30389 | 1870 | 2017 | 2460 | 2899 | 3043 |
| Jamaica | 33.8 | 41.3 | 46.7 | 49.4 | 51.8 | 52.0 | 0.9 | 2.7 | 4.0 | 3.3 | 4.0 | 2.7 | 429 | 752 | 1256 | 1921 | 3479 | 4907 | 1061 | 1068 | 1165 | 1479 | 1052 |
| Jordan | 50.9 | 56.0 | 59.9 | 72.2 | 79.8 | 82.5 | 0.9 | 1.0 | 2.2 | 3.3 | 3.2 | 3.4 | | 424 | 793 | 1312 | 1763 | 4371 | 313 | 698 | 1033 | 1014 | 1175 |
| Kazakhstan | 44.2 | 50.2 | 54.1 | 56.3 | 55.7 | 53.7 | | | | | 8.6 | 15.2 | | | | 1647 | 1229 | 9071 | | | 4493 | 2397 | 4561 |
| Kenya | 7.4 | 10.3 | 15.6 | 16.7 | 19.9 | 23.6 | 0.3 | 0.3 | 0.4 | 0.2 | 0.3 | 0.3 | 98 | 142 | 447 | 366 | 406 | 787 | 453 | 452 | 455 | 449 | 482 |
| Korea, Dem. Rep. | 40.2 | 54.2 | 56.9 | 58.4 | 59.4 | 60.2 | | 1.7 | 7.2 | 12.1 | 3.4 | 2.9 | | | | | | | 1311 | 1748 | 1645 | 863 | 767 |
| Korea, Rep. | 27.7 | 40.7 | 56.7 | 73.8 | 79.6 | 82.9 | | 1.7 | 3.5 | 5.8 | 9.5 | 11.5 | 156 | 276 | 1674 | 6153 | 11347 | 20540 | 516 | 1081 | 2171 | 4003 | 5059 |
| Kuwait | 74.9 | 85.7 | 94.8 | 98.0 | 98.1 | 98.2 | 29.8 | 33.4 | 18.0 | 23.5 | 28.9 | 31.3 | | 3829 | 20882 | 8947 | 19787 | 40091 | 7571 | 7620 | 4423 | 9865 | 10893 |
| Kyrgyz Republic | 34.2 | 37.5 | 38.6 | 37.8 | 35.3 | 35.3 | | | | 0.5 | 0.9 | 1.2 | | | | 609 | 280 | 880 | | | 1705 | 473 | 515 |
| Lebanon | 42.3 | 59.5 | 73.7 | 83.1 | 86.0 | 87.1 | 1.4 | 1.7 | 2.4 | 3.4 | 4.7 | 4.7 | | | | 1050 | 5335 | 8552 | 785 | 949 | 723 | 1517 | 1470 |
| Libya | 27.3 | 49.7 | 70.1 | 75.7 | 76.3 | 77.6 | 0.5 | 15.6 | 8.7 | 8.6 | 9.1 | 9.8 | | | | 6549 | 6785 | | 727 | 2238 | 2622 | 3072 | 3578 |
| Macao SAR, China | 95.3 | 97.0 | 98.5 | 99.8 | 100.0 | 100.0 | 0.3 | 0.8 | 2.1 | 2.9 | 3.8 | 1.9 | | | | 8312 | 14128 | 53046 | | | | 2011 | 2569 |
| Malaysia | 26.6 | 33.5 | 42.0 | 49.8 | 62.0 | 72.0 | | 1.3 | 2.0 | 3.1 | 5.4 | 7.7 | 299 | 392 | 1803 | 2417 | 4005 | 8754 | 545 | 860 | 1183 | 1400 | 2011 |
| Marshall Islands | 35.6 | 53.5 | 58.3 | 65.1 | 68.4 | 71.5 | | | | 1.0 | 1.5 | 2.0 | | | 1177 | 1659 | 2127 | 3113 | | | | | |
| Mauritius | 33.2 | 42.0 | 42.4 | 43.9 | 42.7 | 41.8 | | | 0.6 | 1.4 | 2.3 | 3.2 | | | | 2506 | 3861 | 7587 | | 453 | | | |
| Mexico | 50.8 | 59.0 | 66.3 | 71.4 | 74.7 | 77.8 | 1.6 | 2.2 | 3.8 | 3.7 | 3.7 | 3.8 | 338 | 671 | 2763 | 3052 | 6664 | 8885 | 786 | 1352 | 1423 | | 1518 |
| Micronesia, Fed. Sts. | 22.3 | 24.8 | 25.8 | 25.8 | 22.3 | 22.5 | | | | | 1.3 | 1.0 | | | | 1528 | 2171 | 2838 | | 246 | 281 | 357 | 511 |
| Mongolia | 35.7 | 45.1 | 52.1 | 57.0 | 57.1 | 67.6 | 1.4 | 2.2 | 4.1 | 4.6 | 3.1 | 4.2 | 165 | 247 | 943 | 1037 | 474 | 2286 | 149 | | 1560 | 515 | 712 |
| Morocco | 29.4 | 34.5 | 41.2 | 48.4 | 53.3 | 56.7 | 0.3 | 0.5 | 0.8 | 1.0 | 1.2 | 1.6 | | | 1172 | 1660 | 1276 | 2823 | | | | | |
| Namibia | 17.9 | 22.3 | 25.1 | 27.7 | 32.4 | 37.8 | | | | 0.0 | 0.9 | 1.5 | | | | 2141 | 2059 | 5079 | | | | | |
| New Caledonia | 37.4 | 51.2 | 57.4 | 59.5 | 61.8 | 61.9 | 10.9 | 22.2 | 14.0 | 9.7 | 10.8 | 15.7 | | 3322 | 8269 | 15055 | 12580 | | | | | | |
| Nicaragua | 39.6 | 47.0 | 49.9 | 52.3 | 54.7 | 57.3 | 0.3 | 0.6 | 0.6 | 0.6 | 0.7 | 0.8 | 128 | 324 | 660 | 244 | 1001 | 1475 | 495 | 474 | 489 | 494 | 507 |
| Nigeria | 16.2 | 22.7 | 28.6 | 35.3 | 42.4 | 49.0 | 0.1 | 0.4 | 0.9 | 0.5 | 0.6 | 0.5 | 93 | 224 | 871 | 322 | 378 | 1437 | 628 | 712 | 738 | 737 | 721 |
| Northern Mariana Islands | 51.2 | 70.1 | 86.8 | 89.7 | 90.2 | 91.3 | | | | | | | | | | | | | | | | | |
| Oman | 16.4 | 29.7 | 47.6 | 66.1 | 71.6 | 73.2 | | 0.3 | 5.2 | 6.3 | 10.0 | 20.4 | 80 | 354 | 5182 | 6455 | 9062 | 20984 | 303 | 996 | 2330 | 3687 | 8262 |
| Pakistan | 22.1 | 24.8 | 28.1 | 30.6 | 33.1 | 35.9 | 0.3 | 0.4 | 0.6 | 0.6 | 0.7 | 0.9 | 81 | 169 | 296 | 360 | 514 | 1025 | 280 | 310 | 386 | 445 | 487 |
| Palau | 56.8 | 59.7 | 62.5 | 69.6 | 70.0 | 83.4 | 1.5 | 11.8 | 12.9 | 15.6 | 6.1 | 10.6 | | 666 | 5096 | 8262 | 8262 | 9602 | | | 4869 | | |
| Panama | 41.2 | 47.6 | 50.4 | 53.9 | 65.8 | 74.6 | 0.9 | 1.4 | 1.6 | 1.1 | 1.9 | 2.6 | 366 | 265 | 1915 | 2137 | 3804 | 7229 | 1055 | 710 | 600 | 841 | 1009 |
| Papua New Guinea | 3.7 | 9.8 | 13.0 | 15.0 | 13.2 | 12.4 | 0.1 | 0.3 | 0.5 | 0.5 | 0.5 | 0.5 | 117 | | 792 | 774 | 655 | 1382 | | | | | |
| Paraguay | 35.6 | 37.1 | 41.7 | 48.7 | 55.3 | 61.4 | 0.2 | 0.3 | 0.6 | 0.5 | 0.7 | 0.8 | | 221 | 1280 | 1340 | 1532 | 3101 | 537 | 652 | 723 | 720 | 741 |
| Peru | 46.8 | 57.4 | 64.6 | 68.9 | 73.0 | 76.9 | 0.8 | 1.3 | 1.4 | 1.0 | 1.2 | 2.0 | 252 | 548 | 1192 | 1208 | 2050 | 5386 | 673 | 650 | 447 | 470 | 656 |
| Philippines | 30.3 | 33.0 | 37.5 | 48.6 | 48.0 | 48.6 | 0.3 | 0.7 | 0.8 | 0.7 | 0.9 | 0.9 | 254 | 187 | 685 | 715 | 1043 | 2136 | 416 | 473 | 462 | 513 | 434 |
| Puerto Rico | 44.5 | 58.3 | 66.9 | 72.2 | 94.6 | 98.8 | | | | | | | 718 | 1852 | 4503 | 8653 | 16192 | 26106 | | | | | |
| Qatar | 85.2 | 88.4 | 89.4 | 92.8 | 96.3 | 98.7 | 3.7 | 69.2 | 58.5 | 24.7 | 58.5 | 40.3 | | 2760 | 34990 | 15446 | 29914 | 72773 | 7764 | 14801 | 13696 | 18320 | 16559 |
| Saudi Arabia | 31.3 | 48.7 | 65.9 | 76.6 | 79.8 | 82.1 | 0.7 | 7.8 | 17.2 | 13.4 | 14.7 | 17.0 | | 864 | 16692 | 7206 | 9354 | 19327 | 1214 | 3160 | 3687 | 5030 | 7044 |
| Seychelles | 27.7 | 39.1 | 49.4 | 49.3 | 50.4 | 53.2 | | 0.5 | 1.5 | 1.6 | 7.0 | 7.8 | 288 | 344 | 2288 | 5265 | 7579 | 10805 | | | 537 | | |
| Singapore | 100.0 | 100.0 | 100.0 | 100.0 | 100.0 | 100.0 | 0.8 | 8.8 | 13.0 | 15.4 | 12.2 | 9.0 | 428 | 925 | 4913 | 11845 | 23815 | 42784 | 1292 | 2126 | 3779 | 4641 | 6752 |
| South Africa | 46.6 | 47.8 | 48.4 | 52.0 | 56.9 | 61.5 | 5.6 | 6.8 | 8.3 | 9.5 | 8.4 | 7.0 | 423 | 811 | 2921 | 3182 | 3020 | 7137 | 2010 | 2371 | 2584 | 2483 | 2796 |
| Sri Lanka | 16.4 | 19.5 | 18.8 | 17.2 | 15.7 | 15.0 | 0.2 | 0.3 | 0.2 | 0.2 | 0.5 | 0.6 | 143 | 184 | 273 | 472 | 855 | 2400 | 299 | 308 | 324 | 436 | 477 |
| St. Kitts and Nevis | 27.6 | 34.1 | 35.9 | 34.6 | 32.8 | 31.9 | 0.2 | 0.6 | 1.2 | 1.6 | 2.3 | 4.8 | 242 | 363 | 1111 | 3899 | 9146 | 12904 | | | 529 | | |
| St. Lucia | 21.5 | 23.9 | 26.5 | 29.3 | 28.0 | 18.3 | 0.2 | 0.6 | 1.0 | 1.2 | 2.1 | 2.3 | | | 1131 | 2874 | 4871 | 6814 | | | 405 | | |

| Country | | | | | | | | | | | | | | | | | | | | | | | |
|---|---|---|---|---|---|---|---|---|---|---|---|---|---|---|---|---|---|---|---|---|---|---|---|
| St. Vincent and the Grenadines | 25.9 | 30.7 | 35.9 | 41.4 | 45.2 | 48.9 | 0.1 | 0.3 | 0.4 | 0.8 | 1.5 | 1.9 | 161 | 204 | 600 | 1844 | 3684 | 6173 | | | | | 274 |
| Suriname | 47.3 | 45.9 | 55.0 | 60.0 | 64.9 | 69.3 | 1.5 | 4.3 | 6.5 | 4.5 | 4.6 | 4.5 | 323 | 664 | 2174 | 955 | 1912 | 8321 | | | | | 1367 |
| Swaziland | 3.9 | 9.7 | 17.8 | 22.9 | 22.6 | 21.3 | 0.1 | 0.8 | 0.8 | 0.5 | 1.1 | 0.9 | 100 | 252 | 898 | 1292 | 1433 | 3094 | | | | | 358 |
| Syrian Arab Republic | 36.8 | 43.3 | 46.7 | 48.9 | 51.9 | 55.7 | 0.7 | 1.0 | 2.3 | 3.0 | 3.1 | 2.9 | 187 | 336 | 1458 | 989 | 1180 | | 360 | 499 | 840 | 963 | 1005 |
| Tajikistan | 33.2 | 36.9 | 34.3 | 31.7 | 26.5 | 26.5 | | | | | 0.4 | 0.4 | | | 496 | 139 | | 740 | | | 1002 | 347 | 311 |
| Thailand | 19.7 | 20.9 | 26.8 | 29.4 | 31.1 | 33.7 | 0.1 | 0.4 | 0.8 | 1.7 | 3.0 | 4.4 | 101 | 192 | 683 | 1508 | 1969 | 4803 | 361 | 464 | 741 | 1159 | 1768 |
| Tonga | 17.6 | 20.2 | 21.2 | 22.7 | 23.0 | 23.4 | 0.2 | 0.3 | 0.4 | 0.8 | 1.2 | 1.5 | | | 573 | 1193 | 1925 | 3547 | | | | 266 | |
| Trinidad and Tobago | 17.3 | 11.9 | 10.9 | 8.5 | 10.8 | 13.4 | 3.0 | 9.5 | 15.6 | 13.9 | 19.3 | 38.2 | 631 | 869 | 5746 | 4148 | 6431 | 15562 | 2758 | 3527 | 4900 | 8560 | 16091 |
| Tunisia | 37.5 | 43.5 | 50.6 | 57.9 | 63.4 | 66.1 | 0.4 | 0.7 | 1.5 | 1.6 | 2.1 | 2.5 | | 281 | 1370 | 1507 | 2245 | 4207 | 318 | 512 | 607 | 764 | 917 |
| Turkey | 31.5 | 38.2 | 43.8 | 59.2 | 64.7 | 70.5 | 0.6 | 1.2 | 1.7 | 2.7 | 3.4 | 4.1 | 508 | 491 | 1567 | 2791 | 4220 | 10135 | 549 | 716 | 977 | 1209 | 1457 |
| Turkmenistan | 46.4 | 47.8 | 47.1 | 45.1 | 45.9 | 48.4 | | | | | 7.9 | 10.5 | | | | 881 | 645 | 4393 | | | 4776 | 3304 | 4497 |
| Turks and Caicos Islands | 47.7 | 51.1 | 55.3 | 74.3 | 84.6 | 93.3 | | | | | 0.8 | 5.2 | | | | | | | | | | | |
| United Arab Emirates | 73.5 | 77.7 | 80.7 | 79.1 | 80.2 | 84.0 | 0.1 | 65.9 | 36.4 | 28.8 | 37.2 | 19.9 | | | 42962 | 28066 | 34476 | 34049 | 3718 | 7126 | 11306 | 11216 | 7475 |
| Uruguay | 80.2 | 82.4 | 85.4 | 89.0 | 91.3 | 92.5 | | 1.7 | 2.0 | 1.3 | 1.6 | 2.0 | 490 | 761 | 3486 | 2990 | 6873 | 11520 | 857 | 907 | 724 | 931 | 1238 |
| Uzbekistan | 34.0 | 36.7 | 40.8 | 40.2 | 37.4 | 36.2 | | | | | 4.9 | 3.7 | | | 651 | 558 | | 1377 | | | 2261 | 2059 | 1532 |
| Venezuela, RB | 61.6 | 71.9 | 79.2 | 84.3 | 89.9 | 93.3 | 7.5 | 7.0 | 6.0 | 6.2 | 6.2 | 6.9 | 1136 | 1212 | 4447 | 2382 | 4800 | 13559 | 1764 | 2347 | 2206 | 2312 | 2600 |
| Vietnam | 14.7 | 18.3 | 19.2 | 20.3 | 24.4 | 30.4 | 0.0 | | 0.7 | 0.3 | 0.7 | 1.7 | | | 98 | 433 | | 1334 | 302 | 268 | 271 | 370 | 678 |
| Virgin Islands (U.S.) | 56.5 | 69.6 | 80.1 | 87.7 | 92.6 | 95.3 | | | | | | | 756 | 3476 | 7503 | 15051 | | | | | | | |
| Zimbabwe | 12.6 | 17.4 | 22.4 | 29.0 | 33.8 | 38.1 | | 1.6 | 1.3 | 1.5 | 1.1 | 0.7 | 281 | 362 | 916 | 840 | 535 | 568 | 1010 | 891 | 889 | 789 | 688 |

# Index